Preserving the Fruits of the Earth

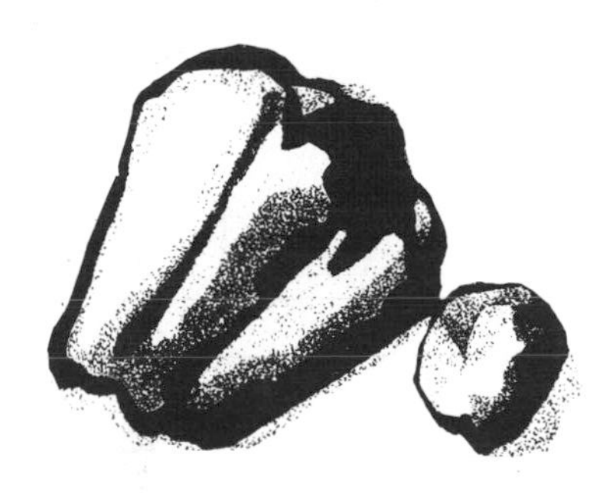

THE DIAL PRESS
NEW YORK
1973

Preserving the Fruits of the Earth

How to "put up" almost every food grown in the United States—in almost every way

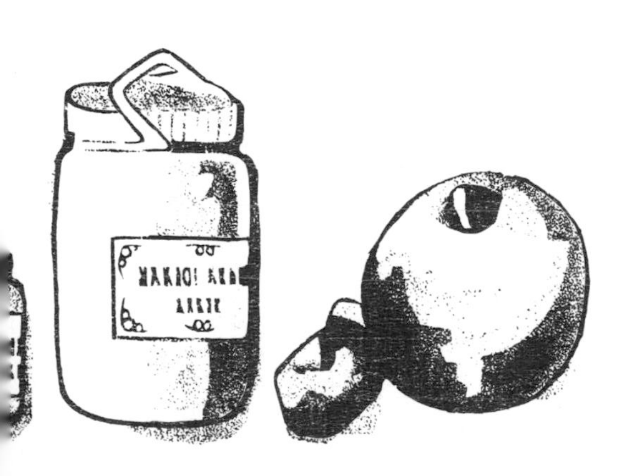

Stanley Schuler and
Elizabeth Meriwether Schuler

Library of Congress Cataloging in Publication Data

Schuler, Stanley.
Preserving the fruits of the earth.

1. Food—Preservation. I. Schuler, Elizabeth
Meriwether, joint author. II. Title
TX601.S34 641.4 72-12816

Printed in the United States of America

Book design by Margaret McCutcheon Wagner

Second Printing, 1973

Contents

Preserving the Fruits of the Earth

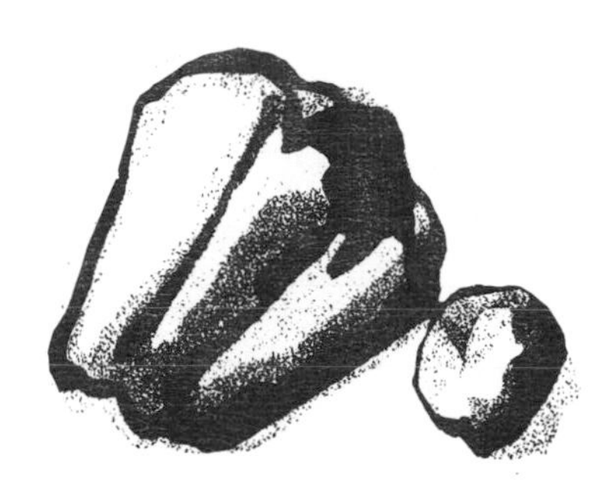

1

The Pleasures of Preserving the Fruits of the Earth

It was long after we had started writing this book—when we were deep in the throes of preserving the summer's bounty—that we found ourselves wondering what had inspired us to write it in the first place.

Our question didn't stem from any sudden loss of interest in the subject. On the contrary, we were more enthusiastic than ever.

What baffled us (in a happy sort of way) was why two people who had been born and raised in large metropolitan areas should be so wrapped up in preserving food that they wanted to write a book about it. Somehow the undertaking just didn't figure.

To be sure, we had, and have, vivid memories of little, old corner grocery stores crammed with large cans, barrels, and tubs from which the grocer doled out such things as salt mackerel, lard, and molasses. We recalled with delight early-morning trips to the Farmers' Market, where you could find every usual and unusual food imaginable—poke and hog's head and scrapple of rare flavor. And we had more than once regaled our daughters with stories about grandfather's vegetable garden, berry-picking expeditions into the woods and fields surrounding his home, and making ice cream on his veranda.

But despite these experiences, the plain fact of the matter was that neither of us had grown up in homes in which foods were put up—or, as some say, put down—in great quantities.

So why, we asked ourselves, do we today spend so much time preserving food when supermarkets are crammed with excellent food in many convenient and long-lasting forms?

The reasons, we eventually concluded, are many; and the chances are that they are pretty much the same reasons most people have for preserving food:

1. It cuts the cost of living. The extent of the saving, of course, depends on whether you raise or catch your own food, or whether you buy it. It also depends on the type of food and on how you preserve it. But it is our observation that, as long as you do not place a high monetary value on the time you put into a preserving project, you can usually put up food for less than you pay in a store.

For example, as we write, one of our two freezers is pretty well filled with fruits and vegetables we grew last summer. If bought in frozen packages at a store, this abundant harvest would have cost approximately $105. By contrast, if we consume all this food within the next eight months—as we almost certainly shall—its actual cost will come to about $50. This figure includes the cost of seeds, fertilizer, spray chemicals, freezer wrap, and electricity; the cost of the freezer amortized over 15 years; and the cost of probable future repairs to the freezer. It does not include the cost of our labor in raising and preserving the food because we consider the work fun and because it was done outside regular working hours.

Many, many thousands of American families can tell a similar story; and recent statistics indicate that many thousands more are hopeful of enjoying the same experience. Witness the following facts:

Sales of chest-type freezers in 1970, a recession year, jumped 19 percent while sales of the more widely used uprights rose 10 percent.

In the same year the Ball Company reported a sharp increase in the number of glass jars it sold for home canning and jelly making.

Also in that year, Northrup, King, the big seed supplier, told us: "The sudden surge in the purchase of vegetable seed packets does not correlate with the much more gradual rise in the number of homes. We have noticed this rise for several years. However, the jump this year has been spectacular."

2. If you hunt or fish, the only way you can enjoy a big kill or a good catch over a period of time is to preserve it. This, of course, has always been true. People in Massachusetts have been salting down cod almost from the time the Pilgrims landed at Plymouth Rock. Indians in Oregon have been smoking and drying salmon even longer.

Today, the number of people who hunt or fish from necessity is small, but the number who hunt or fish for pleasure is soaring. As a consequence, home preserving of fish and game is also soaring.

Sad to say, the resulting delicacies are not always as inexpensive as one might hope. We have acquaintances who recently bagged an elk in Wyoming and had the frozen, butchered carcass air-expressed to their home in Michigan. Not counting their own traveling expenses, they figured the cost of their 300 pounds of meat came to $900. But who's going to quibble about $3-a-pound steaks when he can tell dinner guests the following spring: "Oh, that's the elk we shot in Wyoming last October."

3. By preserving food at home, you can add tremendous variety to your diet. On the other hand, families that do all their food-buying in stores are limited to those foods for which the stores find a big and ready market.

The home preserver can achieve variety in two ways. One is by putting up common foods in uncommon ways. The other is by putting up foods that almost never reach the marketplace. Both categories are large; the second, especially so.

In this book preservation directions are given only for those foods that are cultivated in the United States or are frequently sought out and collected from the wild. We have not included the offbeat wildings to which the delightful Mr. Euell Gibbons has devoted much attention. Yet despite this limitation, more than one-third of the foods discussed in the encyclopedic section are rarely if ever available in stores.

4. Home-preserved food is often of better quality than commercial food. This is not to say that we don't think highly of commercial food. With the exception of an occasional item, it is excellent. But home-preserved food that has been grown at home frequently comes out ahead because there is a shorter time lag between harvesting and processing. It may also be more appealing to the taste buds because it is prepared for the individual rather than the mass palate.

5. Finally, we like—and we think others like—putting up our own food because it is a wonderfully creative activity for families in modern America—one of the many satisfying jobs taken away from us years ago by industry.

Yes, perhaps we're showing off a bit when we send a basket of our own jellies to a friend at Christmastime. It makes us smile to ourselves as, in our imagination, we hear her saying, "Amazing to think that Elizabeth and Stan actually made this themselves! Where did they ever find the time? And you know the jelly's good—really very good."

But desire for applause isn't what makes us do our own preserving. There are two other far stronger goads:

One is a desire to prove to ourselves that we can do what our forefathers did and what industry does—and do it just as well.

The other is a desire to return just a step or two to the simpler life represented by flitches of bacon hanging in the smokehouse, hominy in barrels and the sweet-sour fragrance of pickles cooking on the back burner.

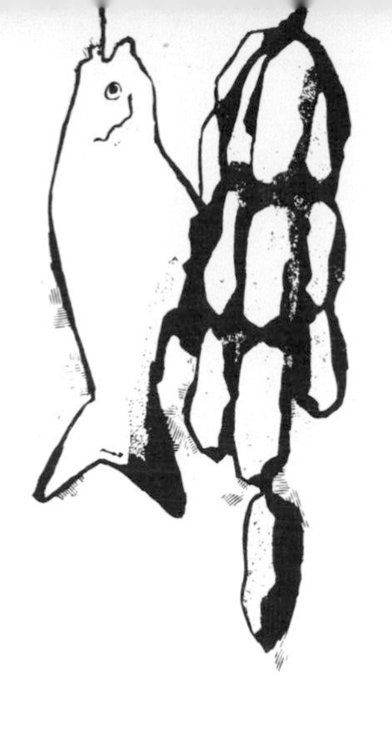

2
How to Dry Foods

If drying was not man's very first method of food preservation, it was certainly one of the first. It's easy to imagine what happened. He had killed a goat or sheep for dinner. When he realized he couldn't immediately consume all he had butchered, he hung the pieces in a tree away from the dogs and forgot about them. Several weeks later he suddenly remembered them. To his disgust, he found that what had been red and juicy had turned brown and leathery in the hot desert air. But being hungry, he stuffed a big piece into his mouth, and as he chewed, the tough, dry mass turned soft and juicy once again. It tasted pretty good. So the next time he killed more meat than he could eat, he hung it in the tree and didn't forget it.

Centuries later the Indians of North America discovered that if they dried lean strips of buffalo meat or venison, ground them into a powder, mixed this with fat, and formed it into thin cakes which they dried in turn, they had a nutritious, reasonably palatable food which kept for months. They called it pemmican.

Today, despite the development of other food preservation methods, dried, or dehydrated, foods are more plentiful than ever before. The great majority, however, are put up by commerical firms. Home processors do little drying except of herbs. This is mainly because the drying of most other foods, though simple, is a rather slow, tedious process. And not everyone is enthusiastic about the flavor and texture of the end product.

Be this as it may, drying is a preservation method you may want to try. In addition to herbs, it is adapted to certain fruits and vegetables, beef, and venison.

How to dry herbs. Herbs that are grown for their seeds require little work. All you have to do is cut off the seed heads as soon as the seeds are dry, shake out the seeds, and pop them into a clean, tight bottle. If you are worried that the seeds will scatter far and wide before you harvest them, cut off the seed heads just before the seeds are fully dry and spread them on screen wire or cheesecloth to finish drying.

Herbs grown for their leaves should be harvested before they flower, because they have more flavor and aroma then than after flowering. Collect the leafy stems on a dry, sunny day as soon as possible after the dew has dried. If they are dusty, dirty, or have been sprayed with an insecticide or fungicide, wash them off quickly in cold water.

One way to dry the herbs is to tie small bunches of stems together at the base and hang them upside down from the ceiling. A better procedure (because drying is faster and more uniform) is to pick the leaves from the stems and spread them out in a single layer on screen wire or cheesecloth placed so that air can circulate freely under and over them. Turn the leaves once or twice a day.

Whichever drying procedure you follow, do it in a dry, airy place out of the sun—a garage or carport, for example.

When the leaves are completely dry and crisp, package them—crushed or whole—in a clean, tight container. They will keep for about a year without serious loss of flavor.

How to dry fruits. In order to dry fruits, you must first build trays to hold them. If you do your drying outdoors, a good size for the trays is 14 by 24 inches. If you dry in your kitchen oven (more about this shortly), make the trays 1 1/2 to 2 inches smaller in length and width than the oven. Construct the frames of 1-inch-square lumber and cover the bottom with strong cloth netting, fiberglass screen cloth, or aluminum screen cloth. Wash and dry the trays thoroughly after every use, and store them in a well-ventilated place to prevent mildewing.

Fruits to be dried should be picked when they are ready to eat. Wash them and dry with paper towels. Cut up large fruits as you like. Remove stems and pits.

Because many fruits turn dark during preparation and storage (this does not spoil them for eating, however), they are usually given special antidarkening treatment first. The favorite method is to expose them to burning sulfur. This not only preserves color as well as flavor but also prevents souring and insect infestation, and reduces loss of vitamins.

Spread the fruits cut side up in a single layer in your trays, and place the trays in a stack outdoors. The bottom tray should be 10 inches off the ground. Separate each tray from the other by 1-inch blocks set under the corners.

Powdered sulfur for treating the fruit is available from hardware and garden supply stores. For fruits that are to be given a 3-hour treatment or less, use 1 teaspoon of sulfur for each pound of fruit. For fruits treated for more than 3 hours, use 3 teaspoons of sulfur. Pour the sulfur in a strip on a small piece of paper, wrap the paper around it and twist the ends. Put the roll in a pan under the stack of fruit trays and light it with a match. Then immediately cover the entire stack with a tight wood or heavy cardboard box. This should be propped up off the ground about 1/2 inch to provide air for combustion.

As soon as the specified sulfuring time is up, the fruit is ready for drying.

Two other less desirable methods of keeping fruits from darkening are to dip them in a solution of 1 gallon of water and 4 to 6 tablespoons of salt, or to blanch them in steam.

Still another treatment, used mainly for figs and sometimes for apricots and pears, is to blanch the fruit in a sirup made of 1 cup of sugar in 3 cups of water. Bring this to a simmer, drop in the fruit, and hold at simmer for 10 minutes. Remove the kettle from the range and let the fruit stand in the sirup for another 10 minutes. Then lift out the fruit, drain it, and let it cool.

The actual drying of fruit can be done outdoors or in a gas or electric oven. Outdoor drying is called sun-drying although the trays of fruit are in reality placed off the ground out of direct sunlight. This method is much slower than oven-drying, and may take as much as a week or 10 days even in the Southwest, where the climate is ideal.

Oven-drying of fruit takes only from 6 to 24 hours, and, for this reason, the dried fruit has better flavor, color, and nutritive value. It is also more likely to be free of insects and insect eggs.

Oven temperatures must be maintained at no more than 165°. This is fairly easily accomplished in the latest model ranges, but you must keep the oven door open in older models and place a thermometer on the upper rack. The top element must also be removed from old electric ranges.

Because most ovens cannot be set at the very low temperatures to which the fruit should first be exposed, preheat the oven at the lowest possible thermostat setting and let it cool before putting in the fruit. From then on you can control the temperature by turning the oven on for brief periods and by opening and closing the oven door.

The fruit should be spread on the trays in a single layer. The lowest tray is placed on an oven rack about 3 inches off the oven floor, and a second tray is set on top of this on 1-inch blocks. Two more trays can be placed on the upper oven rack. Switch the positions of the trays frequently in order to get uniform results.

You can tell by the feel of the fruit whether it is drying properly. If it feels moist and cooler than the oven air, all is well. If it doesn't feel moist and is practically at the same temperature as the oven air, it is drying too fast. To make a more accurate test of drying, remove a piece of fruit from the oven now and then and let it cool. Warm food always feels softer or more moist than cold food.

As you near the end of the drying period, squeeze a handful of fruit. If it leaves no trace of moisture on your skin and if the fruits spring apart when you open your hand, the job is done. Now put the fruit in a tight container for a week to "sweat." This equalizes the moisture content of the pieces and gives them a candylike consistency. They should then be stored in an airtight container in a dark place at normal house temperature or in a refrigerator. Figs are best stored in a freezer.

Check the containers occasionally to make sure the contents are bone dry. If not, they should be reheated for about 30 minutes at 150°, and then resealed.

To prepare dried fruits for use, barely cover them with hot or cold water and let them soak for several hours until they return to their original size and appearance. If they are to be cooked (as they usually are), use the water in which you soaked them. Be careful not to add too much sugar because dried fruit has a higher sugar content than fresh fruit. This results from the fact that the drying process changes the starches in fruit to sugar.

How to dry vegetables. Shell beans and similar vegetables are often left to dry completely on the plant, and after shelling are then placed directly in airtight containers and stored. Sometimes, to prevent weevil infestation, they are heated for a half hour in a 130° oven before packaging.

A better procedure, which preserves more vitamins, better flavor, and truer color, is to harvest the vegetables when they are in prime eating condition. Wash and prepare them as quickly as possible. Then blanch small quantities in boiling water for the time specified in the encyclopedic section of this book under the entry for the individual vegetables. Drain thoroughly and spread in trays in a layer no more than 1/2 inch thick. Dry in your oven until brittle. Then allow them to cool, and store in airtight containers in a dark place at room temperature.

To use, soak 1 cup dried vegetable in 6 cups water for 20 minutes; then bring to a boil and simmer till done. If necessary, you can add a little more water during cooking; but you should try not to add more than the vegetable can soak up.

How to dry meats. See *Venison* in the encyclopedic section.

3

How to Dry-store Foods

You have heard of cold storage, of course. But maybe not of dry storage. Well, there is only one essential point of difference between the two. Cold storage means storage under refrigeration. In dry storage there is no refrigeration; foods are held at whatever temperature the air in the storage area happens to be.

If you are city-born and reared, this may not sound like a very good way to preserve foods. But as your agricultural antecedents could have told you, it works very well for a good many vegetables and fruits.

Unlike other methods of food preservation, dry storage (like cold storage) does not change the taste or appearance of food. In other words, food is not processed in any way. It shouldn't even be washed. You just harvest it and tuck it away.

Another advantage of dry storage is that, once you have built the storage facility, you have no further expense. There is no fuel cost and no packaging cost.

Foods you can store dry. The fact that we earlier mentioned vegetables and fruits as good candidates for dry storage does not mean they are the only foods stored this way. Nuts and grains, for example, are also stored dry. However, when anyone talks about dry storage in the home, the major emphasis is on vegetables and fruits—but only on certain kinds.

We won't waste space here listing all the vegetables and fruits that can and cannot be dry-stored. That information is in the encyclopedic section. But two things must be understood.

First, of those vegetables and fruits that you can dry-store, some varieties perform much better than others. These are normally identified in seed and plant catalogs by such terms as "stores well" or "good winter-keeper." Generally speaking, varieties that mature late in the fall store better than those that mature in early fall or summer. For example, we have an apple tree—a Summer Astrachan—that bears very good fruit early in August. But the fruit has no keeping qualities whatever. Within two weeks after picking, it loses much of its crispness. By contrast, our Russet matures its fruit at least six weeks later, and this keeps well in dry storage for a couple of months.

The second point to note about dry-storing vegetables and fruits is that the way they keep depends in good part on the way they have been grown and harvested. For example, if you let onions stick up high out of the soil, they will become badly sunburned and won't keep well. Similarly, if you scratch the rind of an orange with your thumbnail when you pick it, there is a chance that disease organisms will invade it and shorten its storage life.

Actually, there are no all-encompassing rules for raising vegetables and fruits so that they will keep well in storage. Each plant is something of a law unto itself. But if you are careful to nourish and water all plants properly, protect them from pests, harvest their fruits at the right time, and handle them with care, the chances are that you will be able to store them for a good while.

Storage at high temperatures. Among vegetables and fruits, winter squash, pumpkins, and sweet potatoes are unusual because they keep best at a temperature of 50° to 55°. They even do pretty well at 60°. The storage area should be dark, well ventilated, and somewhat humid.

Unfortunately, it is not easy to find or create such a storage area around the home. In old houses, an unheated basement room probably comes closest to providing the right conditions. But in a house without a basement, you may have to turn off the heat in one of the rooms and keep the door closed and the shades drawn.

Storage at low temperatures. Most vegetables and fruits should be stored in a dark, humid, well-ventilated place at a temperature of 33° to 40°. Such a place is difficult to find around the home, but it is not difficult to develop. A root cellar is the answer.

Root cellars probably derived their name from the fact that so many of the crops stored in them were—and are—root vegetables: carrots, beets, turnips, parsnips, potatoes, for example. But they are also used for cabbages, celery, apples, pears, and other above-ground crops.

The ideal location for a root cellar is either in a corner of your basement or adjacent to your basement, but not under the house. You can then enter it from the basement without going outdoors into the cold and without letting freezing air into the cellar.

An in-the-basement cellar is easy and inexpensive to build because all you have to do is wall off a corner with 2 by 4s covered on one or both sides with gypsum board. Cram the spaces between the studs with fiberglass insulation and cram additional insulation into the spaces between the ceiling joists. The door between the basement and the cellar should be covered on both sides with insulating board. Thus you will keep the heat from the basement and upstairs rooms out of the cellar.

In cold climates, you should also cover the outside walls of the root cellar with insulation to keep out below-freezing cold. This insulation should extend from the ceiling down to below the frost line (the depth to which ground normally freezes in winter). It need not extend all the way to the cellar floor. You can either glue rigid sheets of urethane foam to the concrete walls or erect studs and tack fiberglass blankets in between them.

In order to humidify the cellar, sprinkle the concrete floor with water occasionally.

If you build a root cellar adjacent to your basement, you must, of course, dig a sizable hole in the ground. The width and length depend on how much food you expect to store, but we don't recommend anything less than 8 feet in both directions. The minimum depth should equal your own height plus a couple of inches so you can stand upright; plus 8 to 12 inches for the floor and ceiling; plus the average depth of the frost. In other words, if your frost line is 30 inches deep and if you are 5 feet 9 inches tall, the hole you dig should be a minimum of about 9 1/2 feet deep.

Build the cellar walls of poured concrete or concrete blocks. The latter must be thoroughly waterproofed on the outside. The walls should rest on wide concrete footings and should be built up to within no more than 4 inches of the frost line.

The roof, or ceiling, of the cellar should be a 4-inch-thick, reinforced concrete slab. To make sure that water cannot work down through it, coat the top with asphalt and spread asphalt building paper on this.

How you build the cellar floor depends on the soil drainage. If the drainage is poor, cover the ground with 6 inches of crushed rock and pour a 4-inch reinforced concrete slab over this. When the cellar is in use, sprinkle the floor with water occasionally to humidify the air.

If the soil drainage is good, you don't have to cover the ground at all. Just rake it smooth and pack it down. The alternative is to spread about 2 inches of sand on the ground and lay bricks close together on top without mortar. Both of these floors will allow moisture from the ground to enter the cellar and keep the air humid.

The door from the root cellar into the basement should be insulated on both sides.

No matter which type of root cellar you build, you must provide means for ventilating it so that it does not become too damp and in order to insure that the proper temperature is maintained. This is done by installing two 4- to 6-inch pipes through the side wall of the in-basement root cellar or the roof of the below-ground cellar.

The inlet pipe that brings cold air into the cellar should extend from above the normal snow line down to within about 6 inches of the cellar floor. Cover the top of the pipe with a fixed hood or, better, with a ventilator hood that revolves and brings in air when the wind blows. The outlet pipe that carries off stale, damp, warm air should extend from above the snow line to just below the cellar ceiling. Cover the top with a fixed hood. Both pipes should have dampers so that you can open or close them as necessary to control conditions in the cellar, but some people simply stuff the pipes with rags when it is necessary to close them.

The final step in building a root cellar is to install some simple shelves. These are best made of 1/2- to 1-inch galvanized wire mesh or 2-inch boards spaced 1 inch apart so that air can circulate freely around the stored foods.

Operating a root cellar. The rules are simple:

1. Store sound, unwashed foods in the manner described under the name of the individual item in the encyclopedic section of this book.

2. Check the temperature and humidity of the cellar daily. This is particularly important in very cold and very warm weather. Adjust the ventilating pipes as necessary.

3. Inspect the stored foods frequently, and use or discard any that show signs of deteriorating. By wrapping individual fruits in newspapers, you can discourage the spread of decay from fruit to fruit but you cannot, of course, stop it from ruining individual fruits.

Other methods of dry-storing foods at low temperature. These are all essentially the same. A pit is dug in the ground and lined with something—perhaps a barrel or straw. The vegetables are placed inside, and a large pipe which extends well above ground level is placed in the center of them to provide ventilation. The pit is then covered with straw and enough soil to insulate the vegetables from freezing temperatures.

This is a tried-and-true storage method used by early settlers, the Indians, and perhaps even some of the smarter cave men. It has one serious flaw: Once the soil covering the pit freezes, you need a couple of sticks of dynamite to get the vegetables out.

4
How to Cure and Smoke Meat, Poultry, and Fish

In the oral history of Stan's Mississippi forebears relatively little space is given to his grandfather, Benson Blake. But one short sentence records an event that was undoubtedly considered significant at the time: "Benson, aged four, found a ham while making snares to trip up the Yankee soldiers."

This happened during the siege of Vicksburg. Long before the city fell, Blakely plantation, about ten miles north of the city, was overrun by Yankees looking for plunder. Time and again marauding bands descended on the big house overlooking the Yazoo River; ransacked the rooms and out-buildings; and carried off whatever food, liquor, and valuables they could find. As a result, the defenseless family consisting only of women, children, and a couple of ancient servants was soon on such short rations that Mrs. Blake was reduced to climbing on a rickety buckboard behind a spavined mule and driving into the hills to give General Sherman a piece of her mind.

And then little Benson found a ham. Hurrah!

The evil Yankees had evidently dropped it shortly after cleaning out the Blakely smokehouse. And such was the weight of their haul (or their content of alcohol) that they never noticed the loss. Neither did anyone else until little Benson made his wonderful discovery.

End of simple story.

Yet it's a story we have repeated because it makes a point: If a piece of meat is properly cured and smoked, it will stay edible for an amazing length of time even after lying in an open field under a broiling southern sun.

How to cure meat and poultry. For simplicity's sake, the following material speaks only of meat, but poultry is handled in the same way.

In the South, meat is almost always dry-cured, while in the North it is almost always brine-cured. Both methods result in a good product, but dry-cured meat keeps better without refrigeration.

Meat to be dry-cured must be chilled to below 40°, and during curing it must be held at a temperature of between 35° and 40°. This is about the temperature of your refrigerator's fresh-food compartment. Weigh the meat accurately and figure its curing time carefully. Depending on the thickness of the meat, curing will take 20 to 40 days.

The cure is made of 8 pounds of dairy or kosher salt, 3 pounds of white sugar, and 3 ounces of saltpeter (available at the drugstore). Mix the ingredients together thoroughly (a flour sifter does the job well), and apply the mixture at the rate of 1 1/4 ounces per pound of meat. In the case of hams and other thick cuts apply the mixture in three equal doses: one-third at the start, one-third on the third day, and the final one-third on the tenth day. In the case of bacon and other thin cuts, apply the entire amount at the start.

Application is made by working some of the mixture into the crevices of the meat and covering the outside completely, until it looks as if it had been hit by a heavy frost. Then wrap the meat loosely in aluminum foil or freezer paper, or put it in a crock so that the salt will not be rubbed off during storage, and place it in your refrigerator.

At the end of the curing schedule, soak the meat in cold water to remove surface salt. Thick pieces should then be put back into the refrigerator for about 20 days to give the cure plenty of time to penetrate.

The meat can then be smoked or it can be stored without smoking. If you follow the latter course, remove it from the refrigerator and let the surface dry completely. Then wrap it in the manner described for smoked meat, and store it in any convenient dark, tightly screened, well-ventilated space. It will not spoil, even if the storage space temperature goes as high as 100°. Don't be surprised, however, if the meat loses weight. This is a normal result of dry-curing and subsequent aging.

To brine-cure meat, chill thoroughly, weigh, and put it in a spotlessly clean crock or barrel. If curing several cuts at the same time, pack them in tightly. Cover with a brine made by dissolving 8 pounds salt, 3 pounds sugar, and 3 ounces saltpeter in 4 1/2 gallons boiled water. Let this cool completely before pouring it over the meat. To keep the meat under the liquid, weigh it down under a board topped by a stone.

Put the crock in your refrigerator for the duration of the curing period.

To make sure all the cuts are fully exposed to the brine, you must "overhaul the pack" 7, 14, and 28 days after starting the cure. Simply pour off the brine, take out the meat, repack the crock in a different way, stir the brine, and pour it back over the meat.

If the brine becomes sirupy or stringy looking during the curing process, throw it away, and start over. Scrub the meat in hot water, wash and scald the crock; repack the meat, and cover it with a new cold brine made with 5 1/2 rather than 4 1/2 gallons of water.

When curing is completed, the meat is usually smoked; but it can simply be drained and stored in the refrigerator until used. Unlike dry-cured meat, brine-cured meat gains rather than loses weight in the form of moisture. This moisture shortens storage life at room temperatures.

How to smoke meat. Smoking gives food flavor, changes its color, and dries it out so that it keeps longer. It also slows the development of rancidity and decreases damage from insects.

Probably the simplest smokehouse to build consists of a clean barrel, drum, or even a large cardboard container of the type used by moving men. Remove the two ends and place the barrel upright over the upper end of a sloping trench dug in the earth. The trench should be about 6 inches wide and 6 inches deep. Cover it with boards and a layer of soil. At the lower end of the trench, about 10 feet from the barrel, dig a firepit a couple of feet deep.

To use the smokehouse, run a piece of strong cord or galvanized wire through each piece of meat, and hang the pieces from broomsticks laid across the top of the barrel. It makes no difference how much you fill the barrel as long as the pieces of meat do not touch one another or the sides of the barrel. If the barrel is tall, you can hang one layer of meat above another.

Spread cheesecloth over the barrel so that insects cannot get at the meat, and over this lay a solid wood or metal cover resting on the broomsticks.

The fire in the firepit can be built of any hardwood, but never a softwood. Popular woods are hickory, maple, birch, oak, apple, sassafras, and willow. Corncobs are also very good. Start the fire with shavings of the wood you use, not with paper or softwood.

The main aim in managing the fire is to keep it burning slowly so that it doesn't raise the temperature in the barrel too high. It does not, as a rule, have to give off dense smoke in order to flavor and dry meat.

One way to control the fire is to arrange the sticks like the spokes of a wagon wheel. Lower a sheet of metal down part way over the firepit to reduce the draft. Toss damp sawdust on the flames if they blaze up too high.

Despite its simplicity, our first experience with this type of smokehouse was disappointing because the weather was so cold that the heat of the fire was pretty well dissipated by the time it reached the barrel. So we shifted to an entirely different smokehouse which we have been using ever since even though it looks like a Rube Goldberg invention. In essence, it is nothing more than a big chimney with a two-burner electric hotplate in the bottom.

The base—or heating chamber—is a wooden box 20 inches square by a foot deep. We removed the top and bottom and lined the inner surfaces of the sides with gypsum board to keep them from catching fire. The box is set bottom down on the sandy-gravelly floor of our carport. The hotplate placed inside also rests on the carport floor. We scratched a small opening in the sand under one side of the box to admit air for draft and also to admit the cord of the hotplate. On top of the box is a square of gypsum board with a 14 1/2-inch-diameter circular hole cut in the center.

The smoking chamber above the heating chamber is made of two of the hundred-pound drums in which we buy chlorine for our swimming pool. We removed the tops, cut out the bottoms, placed one drum atop the other, and fastened them together with paper sticky tape and two strips of wood nailed to the sides. The resulting chimney, measuring roughly 15 1/2 inches across and 40 inches high, is placed directly over the hole in the top of the heating chamber. A stainless steel spit taken from a charcoal barbecue is driven through opposite sides of the chimney just below the top. The meat and a weather thermometer hang from this. The top of one of the chlorine drums covers the chimney.

When using this smokehouse, we turn on one of the hotplate burners and place on it a large cast-iron frying pan. In the pan we put chips and chunks of green sassafras, or dead sassafras which has been soaked for hours in water. As the wood smolders, it gives off a pungent smoke which rises up around the meat and pours out of the top of the chimney. In 32° weather, the heat given off by the wood and the hotplate burner is sufficient to raise the temperature of the meat to about 80° if we leave the top off the chimney. To raise the meat temperature, we simply slide the top partway across the chimney; and if this isn't sufficient (though it always has been), we can turn on the second burner. A single layer of inch-thick sassafras chunks requires approximately an hour to burn down to ashes. In order to get at the frying pan when it is necessary to replenish the fire or empty out ashes, we simply lift the smoking chamber with the meat hanging inside off the heating chamber.

It is doubtful if anyone else has built such an odd smokehouse; but we know of many people who have variations of it. Some use old refrigerator shells with holes bored in the sides to provide draft. One acquaintance in Arizona uses an oil drum in which he builds a smoldering fire—without benefit of hotplate—of charcoal and hickory chips.

Regardless of the type of smokehouse you build, all meat and poultry are handled in the same way. When the meat is completely cured, soak it in cold fresh water to remove excess salt. Since it is important not to let the meat become waterlogged, soak it for no more than 5 minutes for each day it has been in cure. In other words, if you cured a cut for 25 days, soak it in fresh water for about 2 hours.

After soaking, wash the meat with hot water and a sharp-bristled brush so it will take on a brighter color. Next, let it dry thoroughly, because you cannot smoke meat that is wet. Then place the meat in the smokehouse, start your fire, and open the ventilators to let out moisture.

When you begin smoking, the smokehouse temperature should be allowed to go just high enough to melt surface grease—about 120°.

But after a couple of hours at this reading, throttle down to about 80° to 100°.

How long you smoke food depends on how much smoke flavor you like. Thirty to forty hours is usually enough. This doesn't mean, however, that you must keep your fire going steadily for 30 to 40 hours. You can let it die out overnight if you wish.

As soon as the meat or poultry has cooled, it should be securely wrapped (unless it is to be aged—see Pork) to keep out insects and to exclude the light and air which hasten rancidity. Wrap each piece separately in wax paper, aluminum foil, saran, or a polyethylene bag; then place it in a tight-fitting cloth bag; and hang it from the ceiling by a long cord so that rodents cannot get at it and air can circulate freely around it.

Smoked, dry-cured meat can be stored in a dark, dry, well-ventilated place at normal room temperatures or below. Smoked, brine-cured meat can be stored in the same way only if it has been smoked for a long time in order to lower its moisture content. Otherwise it must be kept under refrigeration, but because salt lowers the freezing temperature and tends to hasten rancidity, the meat should not be stored for more than a month or two.

During storage, if insect larvae work their way into meat, trim out and burn the infested parts. The sound parts can be eaten any time within three or four days.

Mold which may develop on the outside of dry-cured meat does not affect the wholesomeness of the meat. Just scrub or cut it off.

To test whether a large piece of meat which has been stored for any length of time is edible, stick a long skewer or sharp, thick wire into it along the bone. Insert the skewer from one end of the meat to the center; pull it out, and smell it. Then repeat the process from the opposite end of the meat. If the odor of the skewer is sweet, the meat is sound. An unpleasant odor indicates that the meat has probably spoiled and should be thrown out.

The importance of making this test carefully is indicated by an unpleasant experience we once had when we ran a skewer into a smoked ham from one end only. When withdrawn, it smelled fine, so we went ahead with final cooking. But alas, when the ham was taken from the oven, a large, mushy, oozing spot with a strong odor had appeared. What caused this not even the experts at the state university could decide. The most likely theory was that the ham had been badly bruised during the butchering process and that spoilage had set in around the bone. If we had been more alert when we bought the ham, we might have noticed the bruise and saved the trouble of curing and smoking the meat. But we would almost surely have detected the spoilage if we had skewered the ham properly before cooking it.

How to cure and smoke fish. In the Northwest, the Indians still smoke fish by hanging them in the smoke over an outdoor fire until they are completely dried out. The fish are then stored at normal temperatures in any handy place.

For a tastier product, however, cure the fish you catch and smoke them in your smokehouse.

To prepare the fish, wash and scale them thoroughly as described in Chapter 14. Small fish are cleaned in the usual way and may or may not be beheaded. Large fish are usually filleted. The alternative is to behead the fish and split them down the backbone; remove the entrails but don't cut through the belly skin; then open the fish like a book.

Immerse the fish in a brine for 2 to 4 hours, or even longer if you like a salty product. The purpose of the brine is to give the fish flavor; the preservative effect is minimal. The simplest brine is made with 5 cups salt dissolved in 1 gallon water. Another brine recommended by many sportsmen in Oregon is made of 12 ounces salt, 1 pound brown sugar, and 1 gallon water.

After brining, rinse the fish in fresh water and hang them in a drafty, sunless place to dry for 1 to 3 hours. Then hang them from rods or place them skin side down on wire racks in the smokehouse.

Smoking is done in two ways. Some fish may be smoked by both methods, but most are done by one method or the other.

In the hot-smoking process, smoke fish at about 90° for 12 hours. Then raise the temperature to 150° for 3 hours to cook the fish through. If you cannot maintain this heat, smoke the fish for 3 hours and then put them under the broiler in your range for 10 minutes.

Allow the fish to cool completely. Then wrap them tightly in aluminum foil and freeze. The fish can be stored for 9 to 12 months.

If you prefer to can hot-smoked fish, cut them after smoking into pieces long enough to fit vertically in Mason jars, and pack these in tightly without crushing. Cover with olive oil to 1/2 inch of top. Seal, and process in a pressure canner at 10 pounds pressure. Half-pints for 100 minutes, pints for 110 minutes.

In the cold-smoking process, fish are smoked at about 80°. If you intend to freeze them, expose them to dense smoke for about 6 hours.

For long-term storage without freezing, maintain a fire that gives off very little smoke for the first 24 hours. Then build up a dense smoke and continue smoking for 4 more days or longer until the fish are well dried out. Since the fish have not been cured for a long time in salt, the fire must not be allowed to go out at any time. After smoking is completed, brush melted paraffin all over the fish; then wrap them well in wrapping paper to prevent accidental cracking of the paraffin; and store them in a cool, dry, dark place. The paraffin must be peeled off before the fish are prepared for eating.

How to smoke oysters, clams and mussels. After washing, steam the shellfish briefly until they open. Remove the meats and wash them in cold water. Then place them for 5 minutes in a brine made of 1 1/4 pounds salt to 1 gallon water.

Drain the meats thoroughly, dip them in vegetable oil, and place them on wire mesh in the smokehouse. Smoke at 180° for 15 minutes; turn the meats over, and continue smoking for another 15 minutes. Cool thoroughly; pack in rigid plastic containers, and put them in the freezer. Store for 6 to 8 weeks.

5

How to Salt Fish

In the days before mechanical refrigeration, the most common way of preserving fish—especially large catches—was to salt them. The method still has much to commend it. For one thing, it is applicable to all species of fish, though some are more often salted than others. It is inexpensive. It does not require a great deal of work (but in this respect it cannot be compared with freezing). The fish can be stored for 9 months or longer. And the end product is tasty and interesting.

Brine salting. This is the easier of the two salting processes. The only equipment required, other than a knife for cleaning the fish, is a large, well-washed crock, barrel, or tub.

Small fish are cleaned thoroughly, beheaded, and split down the back. Large fish are filleted. If the fillets are thick, score them lengthwise at 1-inch intervals. The cuts should be 1/4 to 1/2 inch deep.

To remove blood and slime from the cleaned fish or fillets, soak them for 30 minutes in a brine made of 1/2 cup salt in 1 gallon water. Drain well.

Pour a shallow layer of salt into your crock, and pour a deep layer into a wooden box. Dredge each fish in the latter. Take time to rub salt into the body cavity and scored flesh. Then lay the fish skin side down in a smooth layer in the crock and sprinkle a little salt over it. Build up from there, layer upon layer, until you have used up all the fish. Place the top layer of fish skin side up and cover it with a fairly heavy sprinkle of salt. In all, you should use about 1 pound of salt for every 3 pounds of fish.

Cover the fish with a clean, smooth board that fits inside the crock and weight this down with a clean rock. Put the crock in a cool, dark place. Allow small fish to cure for 2 to 4 days, large fish for 5 to 8 days. At the end of this period, remove the fish and scrub them well in a brine made of 2 1/2 pounds salt per gallon of water. Empty and rinse the crock. Then replace the fish in layers with a little salt scattered between them. Make a fresh brine of 2 1/2 pounds salt in 1 gallon water and fill the crock to the top.

Cover the crock with cheesecloth and store it in a cool, dark place. The fish will keep for 9 months but should not be held any longer. Pour off the old brine every 3 months or whenever it shows signs of fermenting, and replace it with fresh brine.

Before serving the salted fish, soak them in cold water for 12 hours or longer. Some people then soak them in milk for several hours.

Dry salting. This is a more tedious process and requires more space. But the finished product is better and keeps longer.

Prepare the fish as for brine salting. After dredging with salt, stack them in a cool, dry, dark place on a rough rack made of laths or boards spaced about 1/4 inch apart. The purpose of the rack is to let the moisture formed by the salting process trickle away. Sprinkle salt on the rack first, and lay the fish skin side down in an even layer with heads and tails alternating. Build up from there, sprinkling salt between layers. The topmost fish are laid skin side up and sprinkled with additional salt. Use roughly 1 pound of salt for every 4 pounds of fish.

Let small fish cure for 2 days, large fish for a week. The stack of fish must be repiled every two days with the top fish at the bottom and the bottom fish at the top. At the end of the curing period, if the weather promises to be damp, let the fish continue curing.

When the fish are ready for drying, wash them well in a brine made of 1/2 cup salt per gallon water. Remove all traces of salt. Then lay the fish skin side down in a single layer on a wire mesh rack under a shed roof. The shed should be open on all sides (except for insect screening) so that the air can circulate freely underneath. The rack should be raised several feet off the ground.

During the first day, turn the fish over four times. Keep them in the shed only during the day. Bring them in at night and also during rainy and cloudy days; pile them in a stack, and weight them down with boards and rocks equaling the weight of the fish. If there are several rainy or cloudy days in a row, repile the fish indoors every other day and sprinkle salt between the layers.

It takes about six clear days to dry large fish, such as cod. To determine whether drying is completed, pinch a thick section of flesh; if this leaves no impression, the fish are done. They should then be wrapped in wax paper or polyethylene film, packed in a tightly covered wooden box, and stored in a cool, dry place. Check them every few weeks to make sure they are holding in good condition. If they look red or moldy, scrub them with brine and dry them outdoors again for a couple of days.

A simpler way to dry small quantities of fish is to tie two or three together at the tail and hang them under the eaves or from the rafters of a shed or carport. But don't forget to bring them in at night and on bad days.

It is also possible to hasten the drying process for any quantity of fish by spreading them on a rack and blowing 80° air over them at high speed with an electric fan-heater.

6
How to Brine Vegetables

Brining is closely akin to pickling but is used simply to preserve vegetables, not to make a pronounced change in their flavor. Vegetables stored in brine in a reasonably cool, dark place can be held for a number of months before they are prepared for table use or put up as pickles.

Although two brining methods are used, one is so troublesome and produces such a salty product that we shall ignore it. The much easier method follows. It is used for snap beans, carrots, cauliflower, celery, corn on the cob, cucumbers, onions, and sweet peppers.

Wash and pick over the vegetables; peel, trim and cut into small pieces. After weighing, put the vegetables in clean crocks or Mason jars. Then make a brine in the proportion of 1 gallon water, 4 1/2 cups salt and 1 pint vinegar, and add at least 1 pint of this to each pound of vegetables.

If you use Mason jars, be sure the vegetables are completely covered by the brine, and cap the jars tightly. If using a crock, weigh down the vegetables with a plate or board and cover the brine with paraffin.

When opening the jars or crock, be sure to note whether any of the vegetables are soft, slimy, or have an unpleasant odor. If so, discard the entire contents of the container.

To prepare the vegetables for serving, parboil in water for 4 minutes to remove excess salt. Drain thoroughly. Then cook in fresh water till done.

7

How to Thresh and Mill Grain

Obviously, if you want to go into milling on a sizable scale, there are numerous steps you can take to set up in business. These range from buying the newest farm and milling machinery to remodeling an ancient mill with a worn millstone in the backwoods of Rhode Island. (We always have had a special admiration for Rhode Island mills because their water-ground corn meal—unlike the white and yellow stuff in the supermarkets—is gray.)

But the odds are that any milling you do—in the immediate future, anyway—is going to be on a minute scale; and that means you will save a lot of money if you do at least part of the job by primitive methods.

Except for corn, the individual grains of the cereals consumed by man are enclosed in hulls and borne in panicles, or spikes, at the top of slender stalks. The grains are harvested when mature and dry by cutting the stalks off at the base with a sickle, scythe, or sicklebar.

The first step in preparing the harvested grain for consumption is to thresh it to remove the hulls, which are collectively known as chaff. There are four ways to do this, all very ancient: (1) Remove the seed spikes from the stalks and rub them between your hands—a slow, inefficient, and rather prickly procedure. (2) Remove the seed spikes, put them in a scooped-out, mortarlike log or rock and pound them with the end of a pestlelike stick. (3) Spread the stalks with their seed spikes on the floor of the barn or garage and walk a horse back and forth over them. (4) Make a flail by attaching a short, heavy stick to a long handle with a leather strip. Then spread the grain on a floor and beat it to pieces.

The grain is now ready to be winnowed. This is another simple process. Just rake away the straw; sweep up the seeds and chaff; and, standing in a windy location over a sheet (or in front of a big electric fan), toss everything into the air. The chaff, being light, blows away; the seeds fall to the sheet.

Bagged in burlap or similar coarse material, the grain can be stored in a dry place for many years.

To mill grain, we recommend that you foresake old-time methods and buy a hand-operated or electric grist mill. All you need do is put the grain in a hopper and turn a handle. The mill can be adjusted for fine, medium, or coarse grains. With it you can turn out a couple of pounds of coarse flour or meal in about 5 minutes. Store this in containers that are tight enough to discourage insects and mice.

8
How to Store Nuts

With some exceptions, the 15 different species of nuts that are grown in the United States are preserved in the same way.

All should be harvested promptly when mature because delay is likely to result in discoloration of the meats; and of course, the squirrels and other animal and bird nuteaters will have extra time to make off with your provender.

Once nuts have been harvested, husk and clean them as necessary, and spread them out in a thin layer in a well-ventilated, sunless place to cure for several days. They can then be stored, in their shells, in mesh bags or equivalent containers in a cool, rather humid place such as a basement. They will keep for a year or more if held at 40°. But even at higher temperatures they will hold well for a long time.

Lacking a proper storage place for whole nuts, the best method of storage is to shell them after they have been cured, package them immediately in moisture-vaporproof containers or polyethylene bags, and store them in your freezer. They will keep almost indefinitely. They may also be stored for a fairly long time in a dark place at normal room temperatures if sealed in tight glass or rigid plastic containers.

Another way to store freshly shelled nuts is to place them in a shallow pan and heat them through in your kitchen oven at 375°. Take care not to roast or parch them. Then pack them into clean Mason jars to 1/4 inch of the top, seal, and process in a boiling water bath—pints for 25 minutes; quarts for 30 minutes. Then let the jars cool on papers, a metal rack, or cutting board for about 12 hours, and store them in a dark, dry place at normal room temperatures or, preferably, below.

Nuts that are roasted, deep fried, or salted have a much shorter storage life than untreated nuts unless they are processed under controlled factory conditions. It is therefore better to wait until just before using nuts to process them.

9

How to Put Up Juice

In the years when we have more fruit than we know what to do with, we make juice from some of it. Fruit juices, like tomato juice, are easy to put up. They taste delicious. And they use up a lot of fruit. In fact, it takes roughly 3 cups of fruit to produce 2 cups of juice.

Start with sound, ripe fruit. Cut out bad spots but don't worry about simple surface blemishes. Wash thoroughly and remove stems; but as a rule, it is not necessary to remove pits, seeds, or cores.

Most—but far from all—fruits are heated to improve juice extraction and inactivate the enzymes that would lower the quality of the juice. The method used sounds ridiculous, but you will be surprised at how well it works.

First drop into 2 quarts of boiling water 10 unscented, white facial tissues. Let these stand for a minute, and beat them into small pieces with a fork. Pour into a strainer and shake out excess water, but do not press.

Now while the tissues continue to drain, place the fruit in a stainless steel or glass kettle and crush it. (Don't use aluminum or galvanized steel equipment, because acid fruit juices pit the former and dissolve the zinc in the latter.) For every 3 parts of fruit add 1 part of the tissue pulp and stir well. Heat no higher than 180°, stirring constantly, until the fruit is soft—no longer. Overheating destroys the fruit flavor.

Pour the fruit and pulp into a jelly bag, set in a colander over a large bowl, and let it drain. The pulp acts as a filter and not only helps to clarify the juice but also prevents clogging of the bag. When the mixture is cool enough to handle, twist the bag to squeeze out the rest of the juice.

If the collected juice is not perfectly clear, strain it through four thicknesses of clean, washed cheesecloth (cheesecloth that has not been washed gives the juice an undesirable flavor). Then mix sugar as desired with the juice until it is completely dissolved. If you are making a blend of two or more juices, this is also the time to mix them together.

Juice is extracted from unheated fruits by a variety of means. Citrus fruits, for example, are squeezed on a reamer. Apples are put through a cider press or electric juice extractor. White grapes are simply crushed and dripped through a jelly bag.

Whether fruits are or are not heated before juice extraction, all except those that are sometimes frozen must now be pasteurized so that they will keep. The best way to do this is in a double boiler, because heating directly over a burner gives juice a cooked taste.

Bring water in the bottom part of the double boiler to a rolling boil. Put the juice, in the upper part of the utensil, over this and bring it up to 190°. Use a jelly or candy thermometer. Stir constantly. Then remove the double boiler from the range.

The juice is packaged in hot, washed, and sterilized canning jars or bottles. If using jars, scald the caps according to the maker's directions. If using bottles, seal them with new, clean, dry crown-type bottle caps.

Fill the jars or bottles to the brim with the hot juice. Work fast, because the temperature of the juice should not drop below 185°. If it does, return the double boiler to the range and heat the juice to 190° again.

Cap the jars or bottles immediately, and turn the jars upside down for 3 minutes, the bottles for 5 minutes. Then immerse them in a large kettle filled with 120° water. After 5 minutes, pour off a third of the water and replace it with cold water from the faucet. Again, after 5 minutes, pour off a third of the water and replace it with cold water. Then, after another 5 minutes, run cold water steadily into the kettle for 5 minutes. By this process you will lower the temperature of the juice to that of the cold water in roughly half an hour.

Finally, dry the containers, label them, and store them in a cool, dry, dark place. The ideal storage temperature is between 32° and

40°. The juice will not spoil at higher temperatures, but it will gradually decline in quality.

The alternative to preserving juice in jars and bottles is to transfer it immediately from the range to your refrigerator. Leave it in the upper part of the double boiler. As soon as juice is cold, pour it into rigid plastic or glass freezer containers. Leave head space of 1/2 inch in pint containers; 1 inch in quart containers. Seal tightly and put the containers in the freezer. The juice will keep in excellent condition for a year.

When using home-preserved juice, make sure that it does not have a disagreeable odor, gas bubbles, or mold.

How to make frozen juice concentrates. After extracting and pasteurizing juice in the manner described above, chill it thoroughly in the refrigerator and pour 3 quarts into a gallon jug. Seal and freeze solid. Then remove the jug from the freezer, open it, and place it upside down on top of a narrow stainless steel, glass, or pottery container inside your refrigerator.

As the concentrated juice thaws, it will drain into the bottom container. As soon as it loses its sweet taste, remove the jug from the container, let the remaining ice thaw, and empty it.

Then pour the concentrated juice back into the jug, freeze it again and let it drain off a second time. Then repeat the process once more, saving the concentrate, and discarding the ice. Finally, pour the concentrate into small freezer cartons and store them in your freezer.

When ready to use a concentrate, dilute it with 3 parts water before serving.

How to put up tomato juice. The process is generally similar to that described above but differs in some details. Follow directions in the encyclopedic section.

How to make fruit sirups. These directions for making strawberry, blackberry, blueberry, raspberry, and grape sirup come from the Oregon Agricultural Extension Service.

Use fully ripe fruit, but make sure—by judicious tasting—that at least half of it is tart. Extract the juice by heating the fruit according to the preceding directions. To make the sirup, combine 1 1/2 cups fruit juice (before pasteurization) with 1 3/4 cups sugar in a large kettle; bring to a rolling boil and boil for 1 minute. Remove from the range, skim, and pour into clean, hot Mason jars. Seal, place in a canning kettle and cover with 1 inch of hot water. Bring to a boil and process for 10 minutes. Then cool, label, and store in a cool, dark place.

The thickness of fruit sirup depends on the kind of fruit and its condition, so it is wise to make a test batch and allow it to cool before putting up all your juice. If the test batch is too thick, let the rest of the juice stand overnight in the refrigerator so that some of the natural pectin will be destroyed. If the test batch is too thin, substitute 1/4 cup white corn sirup and 1 1/2 cups sugar for the sugar called for in the basic recipe. To make a more tart sirup, add 1 tablespoon lemon juice to the basic recipe.

10
How to Candy Fruit

Fruit is candied by impregnating it with sirup ever so slowly until it contains enough sugar not to spoil. The job must be done in such a way that the fruit does not toughen, soften, or turn into jam.

Favorite fruits for candying are figs, peaches, pineapples, and others that are tender when ripe but have a firm flesh. Apples, for example, are too crisp to be candied by the process described here; and most berries are too soft and mushy. The peels of oranges and other citrus fruits are candied, but the method used differs somewhat from the following. Directions for handling these fruits are given in the encyclopedic section under the entries for the fruits.

The first step in candying is to make a sirup in the proportions of 2 cups sweetening and 4 cups water. Sugar may be used but light corn sirup is better because it doesn't dry the fruit so much.

Bring the sirup to a boil and add the washed, prepared fruits; bring to a boil again and cook for 2 minutes. Then remove the kettle from the range and let the fruit stand in the sirup for 24 hours. Use a round board or dish to keep the fruit down in the sirup.

This boiling-standing process is then repeated four times, each time in a heavier sirup. The first sirup is made with 3 cups sweetening in 4 cups water; second, 4 3/4 cups sweetening in 4 cups water; third, 7 cups sweetening in 4 cups water; last, 10 cups sweetening in 4 cups water.

After boiling the fruit for the last time, let it stand in the sirup for 3 weeks. The fruit should be firm and plump. Then remove it from the sirup and dip it quickly in boiling water to remove the stickiness. Place the fruit on wire mesh to dry at room temperature or, better, place it in a 120° oven till dry. When all stickiness has disappeared (you may have to wash the fruit again if the stickiness persists after drying), pack the fruit in shallow containers. Place wax paper or foil between layers. Wrap the containers in plastic film to keep off insects. Store in a cool, dry, dark place. The fruit will keep for a long time.

Another way to handle fruit once it has been candied and dried is to glacé it. Make a sirup of 3 parts sugar, 1 part corn sirup, and 2 parts water, and cook it to 236° on a candy themometer. Then cool to 200°. Dip the fruits in the sirup quickly, drain them on screen wire and allow them to dry. Because the sirup leaves a glaze, or thin sugar coating, the fruit is called glacéd fruit. It is packaged and stored like ordinary candied fruit.

11

How to Make Jelly

In our home, breakfast is the second most important meal of the day and on Sundays it's often the first. Not that we eat huge down-on-the-farm breakfasts. But the day doesn't start out right if we don't have at least fruit, eggs, meat, toast, and jelly or jam, and coffee—six basic ingredients.

Now at first blush you might think that with six basic ingredients to work from, our breakfasts would be infinitely varied. But they're not, and in that respect they are little different from substantial breakfasts that other people eat.

Coffee is coffee. No chance for variety there.

Meat is usually bacon, but sometimes it is sausage, scrapple, or ham. And every once in awhile Stan has fish.

Three or four different breads are used for toast, but since we each eat only two slices a day, it takes a monotonously long time to use up a loaf.

Eggs can be boiled, fried, scrambled, poached, shirred, and made into omelets. That's a pretty good change of pace; but when you get right down to it, an egg is an egg no matter how it is cooked.

Fruit is almost always grapefruit—with orange juice a very occasional change—for about 9 months of the year. In the summer, however, when our garden and orchard are in high gear, we switch around a great deal.

So that leaves the jelly and jam. They are our mainstays for providing variety at breakfast time. By actual count we have found that there are approximately 75 one-fruit jellies or jams made in the United States. And this does not, of course, include the many two and three-fruit combinations that are made. Neither does it include marmalades, preserves, conserves, and butters.

You can see why we like jellies and jams and make lots of them. Oh, no, not 75-plus varieties in one year. If we have just 10 or 15, that's plenty for a change of pace. Going to the jelly closet becomes a pleasurable game. Which will it be this time—strawberry, quince, peach, guava? We hold them up to the light to enjoy their color. We can almost taste them through the glass. What a lovely way to start a new day.

Essentials for jelly. Four things are needed in jelly making: fruit, pectin, acid, and sugar.

Pectin is a natural carbohydrate that causes jellying. All fruits contain it but not in the same amounts. The amount even varies between different varieties of the same fruit, and between fruit picked this year and that picked last year from the same tree. All fruits have more pectin when underripe then fully ripe.

Acid contributes to the flavor of jelly and also to jelling. Here again there is a difference in the amount fruits contain. Ripe fruits always have less acid than underripe fruits.

Sugar also contributes to the flavor of jelly and gel formation. In addition it acts as a preserving agent.

In short, how well you make jelly depends on how well you combine the four essential ingredients. Fortunately, this is not difficult. Anyone can make good jelly—jelly with a pleasing flavor, pretty color, cohesive but delicately quivering figure, and melt-in-the-mouth tenderness.

Equipment. Even though you should make jelly in small batches, you need a big, deep kettle to contain the mixture when it is cooking at a rolling boil. An 8- to 10-quart size is about right. This is also adequate for preparing the fruit.

A jelly bag is needed to drip the fruit. For some reason, we usually wind up making an improvised jelly bag out of three or four thicknesses of cheesecloth. This works perfectly well unless we fill it with too much fruit or squeeze the fruit too hard: then we have a mess on our hands. That's why it is better to sew up an honest-to-goodness bag of unbleached muslin or several layers of cheesecloth.

A jelly, candy, or deep-fat thermometer is useful if you make jelly without added pectin, but it is not essential. Another useful but nonessential gadget is a wide-mouthed funnel for pouring jelly into jars. A tiny double boiler for melting paraffin also falls into the useful-but-not-essential category.

Other equipment needed—spoons, knives, a colander, measuring cups, etc.—is fairly obvious.

Jelly glasses and how to use them. The standard jelly glass holds 8 ounces and has either a screw-on or slip-on cap. But there is no reason why you should not put up jelly in nonstandard jars that are not cracked or chipped.

To prepare glasses, wash them in soapy water and rinse well. Then put them upside down in boiling water and scald them for about 10 minutes. Remove them several minutes before the jelly is done and let them drain upside down. They must still be hot when you fill them.

Lids and bands should be washed and briefly dipped in boiling water.

When you get around to filling your jelly glasses, all you need is a ladle and a steady hand. We also like to use a wide-mouthed funnel because it helps to keep jelly off the rims of the glasses; but as we just said, this is not essential.

If you are using jelly glasses with two-piece screw-on lids, fill them to 1/8 inch of the top. Wipe the rim clean with a paper towel. Put on the disk with the sealing compound next to the glass; and screw the band down tight over this.

If you are sealing jelly glasses with paraffin, melt the paraffin in the top of a double boiler (it may catch on fire if you melt it in a pan directly on the burner). Fill the glasses to 1/2 inch of the top. Wipe the glasses clean on the inside with a paper towel. Then pour a 1/8 inch layer of paraffin on the jelly. This is all you need and is, in fact, better than a thicker layer because it expands and contracts more readily. If bubbles appear in the paraffin, prick them with a knife; otherwise they will leave holes in the top of the congealed paraffin. When the wax is hard, cover the glasses with slip-on metal lids to keep out mice.

Preparing fruit and extracting juice. If you make jelly without commercial pectin, about a fourth of the fruit should be slightly underripe. If you use commercial pectin, all the fruit should be ripe but still firm. In either case, wash the fruit well in cold water but don't let it stand in water.

The method of preparing fruit varies. Follow the directions in the encyclopedic section under the listing for each fruit. Some fruits can be crushed and pressed without heating. Others need to be cooked somewhat, usually in a little water.

When the fruit is ready, pour it into a damp jelly bag which is set in a colander over a large kettle or simply hung over the kettle. To get absolutely clear jelly, let the juice drip until it stops; then either throw away the pulp or use it to make a fruit butter. You will, however, extract more juice by squeezing the bag. In this case, after you have gotten out all the juice possible, re-strain it—without further squeezing—through a couple of layers of damp, washed cheesecloth or a clean, damp jelly bag.

Making jelly without added pectin. This is the old way of making jelly and it is still a very good way provided the fruit is rich in natural pectin. Crabapples, acid apples, sour cherries, and slip-skin grapes are examples of such fruit. You use less sugar than if you add pectin; on the other hand, you must boil the mixture for a longer time and you end up with less jelly.

If you are not sure whether a fruit has enough pectin, make the following test: Pour 1 tablespoon of the cool fruit juice and 1 tablespoon grain or denatured alcohol into a cup, stir slightly, and let it stand for 2 minutes. If a solid mass of jelly forms, the fruit has a high pectin content, and in making jelly, you should use 1 cup sugar for each cup of juice. If several small jellylike pieces form, however, the pectin content of the fruit is only moderate, and you should use only 3/4 cup sugar for each cup of juice.

If the mixture forms many small particles, the fruit has too little pectin to make jelly unless you add commercial pectin.

In any case, don't taste the mixture; throw it down the drain.

If the fruit contains enough pectin, measure it into a large kettle and bring it to a boil. Then add a measured amount of sugar, stir well until dissolved and boil rapidly until the mixture reaches the jellying point.

There are two simple, quick ways to test whether jelly made without added pectin is done. The most common but least dependable way is to dip a cold metal spoon into the boiling mixture, hold it a foot or more above the kettle—out of the steam—and turn it sideways. If the mixture forms two drops that flow together and fall off the spoon in a sheet, the jelly is done.

The second and better test is to use a jelly, candy, or deep-fat thermometer. Before starting to cook your jelly, take the temperature of boiling water (it is not always 212°). Then after boiling the jelly mixture for a while, lower the bulb of the thermometer into it and read the results. When the jelly mixture temperature is 8 degrees higher than the temperature of boiling water, the jelly is done.

(A third way to test jelly for doneness is to pour a tablespoonful onto a cold plate and put it in a refrigerator for 3 or 4 minutes until the mixture jells. This test is a time-waster, however, because you must remove the jelly mixture from the burner while making the test.)

Making jelly with added pectin. This is our favorite method because it is fast and always produces a perfect product as well as a lot of it. No testing of doneness is necessary. The only slight drawbacks are that you use more sugar and you must follow the recipe exactly.

Two types of pectin are sold in grocery stores. One is powdered; the other, liquid. We have always used the latter, and the recipes in the encyclopedic section are based on it. But that doesn't mean there is anything especially good about it or especially bad about the powdered pectin. They make equally good jelly and they are equally easy to use, though the procedure varies somewhat.

To use powdered pectin, measure sugar into a bowl to be added later. Measure the fruit juice into a kettle and mix in one box of pectin. Place over high heat and bring to a hard boil, stirring constantly. Stir in the sugar at once, and bring to a full rolling boil again (a rolling boil is one which cannot be stirred down). Boil hard for 1 or 2 minutes, depending on the pectin manufacturer's directions. Use a clock or timer with a sweep second hand so that you can time the boiling accurately. Stir constantly. Then immediately remove the kettle from the burner, skim off the foam, and pour the jelly into hot, sterilized glasses.

To use liquid pectin, measure the fruit juice and sugar into a kettle and mix well. Place over high heat and bring to a full boil. Stir constantly. Immediately pour in pectin according to directions. Bring back to a rolling boil and boil hard for 1 minute, stirring constantly. Then remove the kettle from the range, skim off the foam and pour the jelly into glasses.

Storing jellies. After filling and sealing jelly glasses, let them stand undisturbed for 12 hours. Then label them and move them into a cool, dry, dark place. The jelly will keep for a long time, but since the quality declines with each passing month, it is best not to overdo matters.

How to make frozen jellies. This is a new way of making jelly out of certain fruits, such as strawberries and grapes, from which you can extract juice without cooking. The principal advantage is that the jelly tastes more like fresh fruit than conventional jelly. A secondary advantage is that you don't have to spend much time over the range. On the other hand, the jelly takes up freezer or refrigerator space which might better be devoted to something else.

Frozen jelly is usually made with added pectin, either powdered or liquid. It is generally felt that powdered pectin gives slightly better color and flavor, but it is more difficult to use. Unfortunately, there is enough variation in the way jellies are made with powdered pectin to prevent us from giving general directions here.

To use liquid pectin, crush ripe fruit, place it in a jelly bag and press out the juice. Measure the juice and sugar into a large bowl and mix well. In a separate bowl, mix pectin with water or lemon juice, as specified, and add it to the fruit juice. Stir for 3 minutes. Then pour the jelly into cool, sterilized jelly glasses or rigid freezer containers to 1/4 inch of the top. Cover tightly (but a perfect seal as with paraffin is unnecessary). Let stand at room temperature until the jelly sets (this may take up to 24 hours). Then store the jelly in your freezer. It will keep for 6 months or a bit longer. Once opened, however, it must be used up quickly.

Frozen jellies can also be stored in a refrigerator, but for only 3 weeks.

12

How to Make Jams, Marmalades, Conserves, Preserves, and Butters

If your dear Aunt Emily gives you a recipe for a delicious fruit concoction that she calls a marmalade, please don't be critical of her if you're convinced it is really a jam. There simply isn't universal agreement about what to call what. Here, for whatever they are worth, are our definitions:

Jam—a thick spread made from and containing crushed, ground, or chopped fruit.

Marmalade—a jam usually containing citrus fruit and rind in small pieces.

Conserve—a jam made of a mixture of fruits plus, as a rule, nuts and raisins. It is usually served as a side dish rather than as a spread for bread.

Preserve—a jam containing whole small fruits or large pieces of fruit, usually of one kind, in a thick sirup. It is served as a side dish or sometimes as a sauce.

Butter—a spiced fruit purée.

All these confections are put up in more or less the same way.

Equipment and packaging. See Chapter 11. Use the same equipment and the same types of jelly glasses.

Making jam, etc., without added pectin. Use a little underripe fruit with the ripe fruit. Wash fruit in cold water and prepare it according to the directions in the encyclopedic section for that particular fruit. Make the pectin test described in the previous chapter. As a rule of thumb, if pectin content is high, use 1 pound sugar for each pound of prepared fruit. For fruit with a moderate pectin content, use 3/4 pound sugar for each pound of prepared fruit.

Measure fruit and sugar into a kettle, bring slowly to a boil, and cook rapidly. Stir frequently to keep the fruit from sticking. To determine when the jam is done, use a thermometer and insert it vertically into the jam at the center of the kettle. When the jam reaches a temperature 9° above that of boiling water, it is done.

The other way to test for doneness is to pour a little jam into a cold saucer and place it in the refrigerator for a few minutes. Take the jam off the burner while making the test. If the mixture in the saucer jells, the jam is ready for packaging.

When the jam is cooked, remove it at once from the range and stir and skim it off and on for 5 minutes. This will help to keep fruit from floating in the finished product. Then ladle the jam into hot, sterilized glasses and seal.

Making jam, etc., with added pectin. If you use powdered pectin, measure sugar into a bowl to be added later. Prepare and measure the fruit into a large kettle, and mix in one package of pectin. Put over high heat and bring to a boil. Stir in sugar at once. Bring to a rolling boil again and boil hard, stirring constantly, for the time required by the pectin manufacturer. Then remove the kettle from the range and stir and skim the jam for 5 minutes before pouring it into your glasses.

If you use liquid pectin, measure prepared fruit and sugar into a kettle and mix thoroughly. Bring to a rolling boil and boil for 1 minute, stirring constantly. Then remove the kettle from the range and immediately pour in the pectin as called for. Skim and stir the jam for 5 minutes before ladling it into glasses.

Making frozen jam. Follow the directions for making frozen jelly in the preceding chapter.

Making fruit butters. If you wish, you can make a good butter from the pulp that is left after extracting juice for jelly. You will get a richer product, however, if you use the entire fruit.

After washing and cutting up or crushing the fruit, cook it slowly with sugar until soft; then put it through a food mill, and continue cooking until done. Doneness is indicated when no rim of liquid separates from a dab of the fruit butter placed on a cold plate. The butter may then be poured into hot, sterilized glasses and sealed. However, since butters are often made with little sugar, it is advisable to pour them into Mason jars with screw-on tops. These are then processed for 10 minutes in a boiling water bath.

Butters are also made with added pectin in the same way that jam is made.

Making preserves. The main difference between preserves and the other concoctions discussed in this chapter is that, after the fruits are cooked with sugar for a brief period, they are usually taken from the range and allowed to stand for several hours before the cooking is completed. This improves the color and texture of the fruit.

13

How to Make Pickles, Relishes, and Chutneys

Just what it is that turns pickle makers into pickle-making addicts we are not sure. But there's something about the process or the end result that does exactly that. We know people who wouldn't dream of preserving food by any other method but who go out of their way to make pickles. There's a banker who, come every August, puts up a dozen jars of corn relish. And a retired purchasing agent who makes catsup by a recipe handed down from his grandmother. And a Sicilian hairdresser who brines olives.

And then there is our youngest daughter, Cary. Ever since she was a child she has absolutely loathed onions, and yet she has been making pickles full of onions since she was in grammar school. One year she even made the mistake of cutting a red hot chile pepper into strips and then pushing back her hair with her fingers. Instantly her head was set on fire, so to speak, by the peppery juice, and it was several hours before we managed to relieve her very real agony with cold compresses. Yet the next day, she was back making pickles again.

"I like them," she says. "I like the aroma when they're cooking. And I love to eat them."

Maybe that's the only explanation there is for pickle-making addicts. Anyway, it will do, because that's exactly what we think.

Pickles can be categorized in various ways: fruit pickles and vegetable pickles (and also fish pickles, but since there are not many of them, they are discussed only in the encyclopedic section). Sweet pickles and sour pickles. Quick-process pickles and long-process pickles. Pickles packed by the open-kettle method of canning and those packed by the hot-water-bath method of canning.

In other words, pickle making is not a uniform food-preservation process producing uniform results; yet there are a number of basic rules that apply to the making of almost all pickles.

Equipment. Kettles in which pickles are made should be constructed of stainless steel, unchipped enamelware, aluminum, or glass. Other materials are affected by the acids or salts in the pickles and are likely to spoil the pickles.

Crocks or jars for fermenting pickles should be of stoneware, pottery, glass, unchipped enamelware or sturdy plastic.

A large kettle of any material is needed for processing pickles by the hot-water-bath method. If you make pickles in any quantity, a 7- to 9-quart canning kettle exactly like that used in canning fruits and acid vegetables is just the ticket. Or you can use a pressure canner if you leave the vent open so that pressure will not build up.

Pickle packing. The kind of jars used to package pickles depends on how they are processed. If pickles are processed by the open-kettle method, any strong glass container such as a Mason jar, jelly glass, or old mayonnaise bottle will do. Wash the containers in soapy water; rinse well; then put them in a kettle, cover them with water, and sterilize them in boiling water for 10 minutes. Fill the containers with the pickles while they are hot. If you are using containers without tight-fitting lids, you should then cover the pickles with a layer of melted paraffin. However, if you are using Mason jars with two-piece, self-sealing caps similar to those used in canning, boil the caps briefly with the jars; fill the jars to the brim with pickles; wipe the rims; and screw on the caps. (Do not provide head space between the pickles and the lid, because the air may darken the pickles.)

Pickles that are processed in a hot-water bath should be packed only in Mason jars with two-piece caps. Wash the jars thoroughly in hot water but don't bother to sterilize them because this will be done in the canning kettle. Fill the jars while they are hot, leaving a 1/2-inch head space; wipe the rims clean; and screw on the caps as tightly as possible.

Pickle ingredients. Vegetables used in pickles should be tender; fruits should be firm and slightly underripe. Process both vegetables and fruits within 24 hours of harvest. This is especially important in the case of cucumbers, because they deteriorate rather rapidly unless stored in a refrigerator. Wash the fruits and vegetables carefully under cold running water; pick off cucumber blossoms; pare and cut up as necessary.

Use pure, granulated dairy, pickling, or kosher salt. Don't use iodized salt because it may darken pickles. Flake salt may be used but, if so, you must increase the amount called for in the encyclopedic section by 50 percent.

Either cider or white vinegar may be used. The former imparts a somewhat mellower, better flavor, but the latter changes the color of pickles less, if at all, and is generally favored for pickling onions, pears, and cauliflower. In either case, use a high-grade vinegar bearing a label which states that it contains between 4 and 6 percent (or 40 and 60 grains) acetic acid. (Incidentally, if you ever made pickles according to one of your grandmother's recipes and were disappointed to find that it was very sour, it is because vinegar made today contains more acetic acid than in the past.)

Sugar for pickles can be either granulated, white cane or beet sugar. Brown sugar should usually not be used.

Unless ground spices are called for in recipes, use whole spices, which hold their flavor better. As a rule, the spices should be loosely tied in a piece of cloth or a cloth bag, and should be removed from the pickle before it is packed. If left in the pickle, they will darken it.

To make pickles crisp, vegetables are sometimes presoaked in lime. If this is called for, use calcium hydroxide (slaked lime), available at a drugstore. Alum may also be designated for the same purpose, but most modern pickle makers frown on it because it is difficult to find and measure, and may actually spoil a pickle.

Open-kettle-method packing. This method is used only for pickles containing enough vinegar or salt to kill any harmful bacteria in the pickles. The procedure followed varies with the pickle. For example, for some pickles all the ingredients are cooked together for a while in an open kettle and are then immediately packed in jars and sealed. In other cases, the principal solid ingredients are soaked overnight in a vinegar-spice mixture; then the whole thing is cooked, packed, and sealed.

Follow directions in the encyclopedic section.

Hot-water-bath processing. This method is always used when the pickles contain insufficient vinegar or salt to kill harmful bacteria. Many home economists—notably those with the Department of

Agriculture—also recommend this method for *all* pickles on the ground that there is always danger of spoilage organisms getting into the pickle when it is transferred from kettle to jar.

The procedure followed varies slightly depending on the pickle and whether it is packed hot, or packed cold and then covered with boiling liquid. In general, however, you place the ingredients in the jars to within 1/2 inch of the top; eliminate air bubbles by running a knife around the inside of the jars; clean the jar rims; and apply the caps tightly. Place the jars in your canning kettle and cover them with 1 to 2 inches of water. When the water comes to a boil, process for the time specified in the encyclopedic section. (But note that the times given are for people living at less than a 1,000-foot altitude. If you live at a higher altitude increase the processing time in accordance with the table in Chapter 15.)

As soon as the processing time is up, lift out the jars and let them cool in a draft-free place on a wire rack, wooden cutting board, or a pad of newspapers.

How to make long-process pickles. Also called fermented or brined pickles, long-process pickles are cured in brine for several weeks before they are processed. Cucumbers are most often used, although other vegetables can be used. The process changes the color of cucumbers from green to olive and makes the interior tender, firm, and translucent.

After the vegetable is washed and dried, it is packed into a crock or large jar and covered with a brine made of salt and water and sometimes vinegar. To keep the vegetables under the brine, weigh them down with a plate topped by some heavy object.

The temperature of the room in which the curing is done should be between 70° and 80°, but can be somewhat lower. During the curing process, you must add salt frequently to maintain the strength of the brine. Remove daily the scum that forms on the brine.

When the curing has been completed, pack the vegetables into small jars and process them in a boiling water bath.

Storing pickles. Once pickles are processed, store them in a dark, dry, reasonably cool place. They will keep for many months.

When opening a pickle jar, be on the lookout for molds, unpleasant odors, changes of color, or a mushy or slippery appearance. If there is any sign of spoilage, don't taste the pickle. Dispose of it at once.

14
How to Freeze Foods

We have been freezing addicts ever since World War II, when a locker plant was established in the town where we lived. After the war we switched from the public locker to an 8-cubic-foot chest freezer at home. And today we have two upright freezers and a combination refrigerator-freezer giving us a total of 33 cubic feet of storage space. Since 1 cubic foot is supposed to hold roughly 35 pounds of frozen food, that means that when the freezers are full—as they often are—we have some 1,150 pounds of assorted foods stored away just waiting to be eaten. Some of these are prepared foods such as pies and breads, but most are vegetables and fruits from our own garden; meats, poultry, and fish from the store.

As opposed to other types of food preservation, we like food freezing for five reasons:

1. It's simple and quick.
2. The method is adaptable to a great many foods.
3. When cooked and/or served, frozen foods more nearly resemble the fresh produce than any other type of preserved foods except those held in dry storage.
4. The nutritive values of frozen foods are high.
5. There is little to go wrong in the freezing process that could result in foods unsafe to eat.

One possible drawback of food freezing is the cost. Because a freezer is rather expensive to buy and operate, the cost of freezing runs higher than that of most other preservation methods (but not higher than the cost of buying food in a store). The Department of Agriculture figures that if you purchase a 15-cubic-foot freezer for

$300 and have an average electrical rate of 1.95 cents per kilowatt-hour, the monthly cost of freezer operation—not counting the cost of the food—is as follows:

Depreciation (based on a freezer life of 15 years)	$1.67
Repairs (2 percent of purchase price per year)	.50
Electricity	1.99
Packaging (2.5 cents per pound of frozen food)	.71
Total	$4.87

This means that if you keep a 15-cubic-foot freezer three-quarters full at all times, each pound of food costs about 15 cents for processing and storage. Comparable figures are not available for home-canned foods, but they are obviously lower because there is no storage cost.

In defense of the freezing process, however, it should be noted that only an infinitesimal percentage of homemakers who do home freezing complain about the cost.

Types of freezers. The most common type of freezer in use today is the combination refrigerator-freezer, a two-door appliance with a fresh-food storage compartment and a zero-degree freezing compartment. In some models the freezer is above the fresh-food compartment; in others it is below; in still others it is on the right or left side. Whatever the design, the freezing compartment works just like a separate food freezer and is just as efficient. The main difference is that the compartment is usually small. (It's possible to find a combination unit with a 16-cubic-foot freezing compartment, but in the vast majority of combinations the compartment runs between 2 and 5 cubic feet.)

The true food freezer is a single-compartment appliance designed as a top-opening chest or as a front-opening upright. Since there is no difference in efficiency or operation and very little difference in the range of sizes available (the largest unit has a capacity of about 30 cubic feet), the choice depends mainly on cost and where the freezer is to be installed.

Chests are less expensive than uprights, though not by a great deal. Uprights, on the other hand, take up less floor space (but not so much as you might think, because you must allow space for the door to swing open) and a great deal less wall space.

Some appliance dealers make the claim that uprights are easier to use than chests because you can get at the stored packages more easily, but anyone who has used both types of freezer knows there is really little difference on this score. Packages at the back of an upright's shelves are just about as awkward to get at as those in the bottom of a chest. What's more, packages at the front of the shelves have a very irritating tendency to fall out on the floor.

Operating a freezer. Whatever the design, freezer operation is extremely simple. Just plug it into a 20-amp, 120-volt circuit which is not used for anything else. Set the control dial to a mid-point and after the box has run for 24 hours, check the temperature with a thermometer. The temperature should be maintained at all times at 0° to minus 5°. Readjust the control dial accordingly.

When putting food into the freezer to be frozen, try to place it directly against the freezer plates or coils, because the faster food freezes, the better. Leave a little space between packages. Don't try to freeze too much food at one time. As a rule, you should not freeze more than 2 to 3 pounds of food per cubic foot of storage capacity.

The food should be frozen hard within 24 hours. After that the packages can be stacked close together anywhere in the freezer.

Defrost the freezer when the frost is 1/4 inch thick. In a manual-defrost model, turn off the current, remove the food to a refrigerator, and leave the freezer door open. The ice will melt rapidly. You can hurry the process, however, by placing pans of hot water inside and scraping off the loose frost with a dull metal spatula or plastic scraper.

Some freezers have automatic defrosting devices or form no frost at all. These, of course, elimate a tiresome job; on the other hand, the cost of operation is higher and breakdowns tend to be more frequent.

After the freezer is defrosted, wash the interior with mild detergent and water, rinse and dry. Even freezers that defrost themselves must be washed a couple of times a year.

The refrigerating mechanism under and behind the freezer should be cleaned at least once a year with a vacuum cleaner to remove accumulated dust, lint, etc.

Should the power fail, there is generally little to worry about. Most outages today last only a few hours, whereas a good freezer is sufficiently insulated to keep foods frozen for at least 24 hours and usually longer, provided you don't open the door very often. If there is reason to think the power will be off for more than 24 hours, try to find a supply of dry ice and fill the freezer with about 3 pounds per cubic foot of storage space. The alternative is to transfer the frozen food to a friend's freezer or a locker plant.

Frozen foods that are thawed—either as a result of a power failure or for some other reason—can be safely refrozen if they still contain ice crystals. But remember that no matter how little frozen foods thaw, their quality is reduced when refrozen. So it is better to use thawed and partially thawed foods at once or as soon as possible after refreezing.

Packaging materials. To maintain the quality of the foods you freeze, you must package them in containers or wrappings that will prevent the escape of moisture and the resultant rapid drying out of the foods. Needless to say, the packaging must also be leakproof.

Rigid and semirigid containers of polyethylene, tempered glass, aluminum, and heavily waxed cardboard are the types most commonly used for foods that are cut or ground up—vegetables and fruits, for instance. Of these, we have found square polyethylene boxes with slightly tapered sides to be the best. They are relatively inexpensive and reusable indefinitely. They are strong, damage-resistant, and hold their shape, especially the heavier grades. They have wide mouths that are easy to fill, and although the slip-on lids sometimes don't slip on as easily as you might like, they give a tight seal once in place. They are easy to wash—you can even do them in a dishwasher. In the freezer, you can set them close together and pile one on top of another; and when not in use, they nest easily and take up little space.

Odd-shaped foods such as meat, poultry, fish, corn on the cob, and asparagus are usually packaged in polyethylene bags or sheets of saran, aluminum foil, or so-called freezer wrap made of paper.

Polyethylene bags are the most pleasant to use because you just slip the food into them, press out the air, and seal. Sealing can be done by wrapping a sheet of thin paper over the open edges and pressing with a warm iron. But a much simpler, and just as reliable, method is to twist the top of the bag tightly and tie it with one of the wire-reinforced ties that come in the box of bags.

Sheet materials are considerably less expensive than bags, although by the time most people get through wrapping a package, they have used more sheet than necessary and have thus reduced its cost advantage considerably. Nevertheless, sheets are needed for wrapping very large food parcels, such as a turkey or whole salmon; and they make much neater packages than bags when you are wrapping small parcels of compact shape (a bunch of leeks, for example).

Of the sheet materials, aluminum foil is the best, provided it is heavy-duty foil made specifically for freezing. Ordinary household foil is too light and tearable. Foil gives long-time protection of foods; it's easy to wrap around most parcels (except whole chickens and other awkward items); and it is often reusable.

Packaging procedures. These are not difficult.

Pack foods tightly to eliminate as much air as possible in the package.

In rigid containers, leave space at the top for the food to expand. For foods packed in liquid, leave at least 1/2 inch in a pint container; 1 inch in a quart container. For dry-packed foods, leave 1/2 inch in all containers.

When using sheet wrappings, tear off a piece long enough to go around the food with plenty to spare. Place the food in the center of the sheet. Bring together the ends of the sheet paralleling the food, inside surface to inside surface, and fold them together, over and over, down tight on the food. Then make a double fold in the other ends of the sheet and fold them under the food. If you are using paper freezer wrap, seal the ends with gummed tape.

If wrapping meat, fish, or poultry with sharp bones that may pierce the wrapping, use two thicknesses of wrapping.

Labeling. In the Schuler household the conversation goes like this:

Elizabeth: "You said you wanted sorrel soup for lunch. But where is it?"

Stan: "Right there on the second shelf of the vegetable freezer."

Elizabeth: "I don't see anything labeled sorrel soup. Is this it?"

Stan: "That looks more like applesauce."

Elizabeth: "When will you ever remember to label things you freeze?"

She's right—ever so right. If you don't want to waste time rummaging through a freezer, opening and examining packages that should be kept closed, you must label everything you store away. Include the name of the food and also the month and year it was frozen. The latter information is important because there is a fairly definite limit beyond which frozen foods should not be kept. (See the encyclopedic section for the limits for various foods.)

All kinds of labels and labeling devices are recommended for frozen-food packages, but none of these is very satisfactory, especially on rigid polyethylene containers. By far the best thing we have found is ordinary brown masking tape of the type used by painters. It sticks to any dry surface and is easy to write on with pencil or pen.

How to freeze vegetables. At the outset, please note that salad greens, radishes, and green onions do not freeze well, so don't waste your time even thinking about trying it. All other popular vegetables, however, freeze well to superlatively. But even here, if you want the very best product obtainable, you should remember that some varieties freeze better than others. We name some, but not all, of these in the encyclopedic section.

Whatever the vegetables you freeze, they should be young and tender. If possible, pick them just before you freeze them. The less time from garden to freezer, the tastier and more nutritious the end product.

Except for those in pods or husks, vegetables should first be washed in cold water. Pick them over carefully. Sort according to size if they are not to be cut into uniform pieces. Trim, peel, and otherwise prepare for freezing as called for in the encyclopedic section.

Blanching followed by immediate chilling are two of the most important steps in freezing vegetables. Blanching, or scalding, inactivates the enzymes in the vegetables and thus prevents deterioration during storage. Chilling is necessary to stop the vegetables from cooking longer than they should.

The best way to blanch most vegetables is by boiling. Just pour a gallon of water into a large kettle and bring it to a boil. Then drop in the vegetables. When the water starts to bubble again, boil them for the exact time called for.

This sounds easy—and is—but several points warrant discussion.

1. Don't blanch much more than 1 pound of vegetables at a time. You can do a lot less, but try not to do a lot more because the results are less uniform.

2. Most people prefer to keep the vegetables in a wire strainer throughout the blanching process because this simplifies handling. However, the strainer should be flat-bottomed—not round-bottomed—because the vegetables can spread out in a thin layer and there is little chance of undercooking some of them.

But there is no rule that you must use a strainer at all. If you want to toss the vegetables loose into the boiling water, fine; they will blanch just as well as in a strainer. The only drawback is that you must then carry the heavy kettle to the sink and pour the contents through a strainer. This is a steamy, sometimes messy, operation.

3. Some home-freezing pamphlets, notably the one published by the Department of Agriculture, recommend that the blanching process be timed from the instant the vegetables are placed in boiling water. As a result, the timing they call for is a little longer than the timing we recommend.

We call this to your attention only to prevent confusion if you happen to pick up a copy of one of these pamphlets. Actually, one process works as well as the other. We prefer ours (which is also recommended by other people) because we're not convinced that everyone carefully weighs out 1 pound of vegetables before blanching; consequently there is less chance of undercooking a load that's too big.

4. If you live more than 5,000 feet above sea level, add 1 minute to the blanching times specified.

5. If you are freezing more than one batch of the same vegetable,

there's no reason why you shouldn't keep on using the same boiling water until it gets scummy.

Other methods of blanching vegetables are by steaming and baking. Baking has some advantage over boiling when you put up winter squash, pumpkins, and sweet potatoes. Steaming does not.

But to get back to the freezing process—

As soon as you have boiled the vegetables for the specified time, take them to the sink, put them in a strainer in a large pot of cold water, and run cold water over them until there is no warmth left in them. This should take about as much time as the blanching process, perhaps a minute or two longer. If you want to speed the process, add ice to the water.

Now let the vegetables drain for a few minutes. Then pack them into clean containers; seal tight; label, and pop them into the freezer. The job is done.

Weeks or months later, when you're ready to eat a frozen vegetable, it will take just a few minutes to get it on the table. The only vegetable that needs to be thawed—and that only partially—is corn on the cob. The process normally followed is to bring about 1/2 cup of water to a boil in a saucepan. Drop in the frozen vegetable, cover the pan, and bring the water to a boil again. Break up the vegetable as necessary with a fork. When the water is boiling throughout the pan, lower the heat and cook gently for the time specified. (Add extra time if you live at a high altitude.) Drain, season, and serve.

If you prefer cooking in a pressure saucepan, follow the cooking times specified by the saucepan manufacturer.

Many frozen vegetables may also be baked or pan-fried.

How to freeze fruits. The trouble with freezing fruits is that fruits are usually eaten fresh, and there is considerable difference between fresh and frozen fruit. Consequently, you may be a bit disappointed with the frozen product.

But that isn't really being fair. If you will remember that processing changes any food, you will have to admit that frozen fruits are excellent—just a little different.

Most fruits freeze quite well. But as is the case with vegetables, some varieties of certain fruits freeze better than others. Also as in the case of vegetables, the best frozen fruits are those that are picked at the peak of flavor and ripeness and gotten into the freezer soon thereafter.

Pick over the fruits carefully and discard those that are not perfect (or cut out the bad spots). Wash well, taking care not to bruise the fruit. Then peel, pit, slice, etc., according to directions in the encyclopedic section. Prepare only a couple of quarts at a time. You can use any kind of utensil except those made of galvanized metal, because the fruit acids dissolve the zinc, which is poisonous. Chipped enamelware and old tinware should also be avoided because the metal may give the fruit a metallic flavor.

Three packing methods are used:

1. In sugar sirup which is made by dissolving granulated cane or beet sugar in boiling water and then cooling to 70° or less before using. How sweet you make the sirup is a matter of taste. It also depends, of course, on how sweet or tart the fruit is. In other words, you don't have to follow our recommendations. Neither do you have to be consistent from one batch to the next.

The table below gives the formulas for sirups of different strengths (see the encyclopedic section for their use with different fruits):

Type of sirup	Cups of sugar	Cups of water	Sirup yield (cups)
30%	2	4	5
35%	2 ½	4	5 ⅓
40%	3	4	5 ½
50%	4 ¾	4	6 ½
60%	7	4	7 ¾

When packing fruit in sirup, first fill the container about one-third of the way with sirup. Pack in the fruit to within 1/2 or 1 inch of the top, depending on the size of the container. Then add enough sirup to cover the fruit. To keep the fruit from floating, crush a piece of plastic film and press it down on the fruit.

2. In sugar. All you do is mix fruit and sugar together in the amounts called for. The consistency of the end product depends on how soft the fruit is to start with and on how vigorously you stir it. A certain amount of juice in the bottom of the mixing bowl is almost

inevitable; and if you are working with very ripe strawberries or raspberries, you will wind up with a great deal of juice and rather mushy fruit.

The only sure way to keep fruit whole and dry is to spread it out on a cookie sheet and freeze it before mixing with sugar. The alternative—less good but also less work—is to place a shallow layer of fruit in the bottom of a flat-bottomed bowl, sprinkle it with sugar and build up from there, alternating fruit and sugar. Then mix the entire batch together very lightly.

3. Unsweetened. You just place the prepared fruit in containers without sugar or sirup. If you're packing rhubarb, figs, gooseberries, cranberries, or currants, this method yields a very satisfactory product. It is not recommended for most other fruits, however, because they need sweetening to preserve quality.

To prevent peaches, plums, apricots, sweet cherries, figs, and other fruits from darkening when frozen, add ascorbic acid to the sugar or sirup. Theoretically ascorbic acid in any form will do the trick, but in actuality that isn't so. We have tried every possible way of dissolving ascorbic acid tablets from the drugstore, and have not succeeded yet. You must use ascorbic acid crystals or powder made specifically for the purpose. (These are usually sold under trade names such as Fruit-Fresh or ACM.) As a general rule, you should use these at the rate of 1 teaspoon per cup of sugar sirup or dry sugar.

If frozen fruits are to be served fresh, they should be thawed just enough to melt almost all the ice crystals. Then serve at once, while the fruits are still slightly crystalline and cold. If they thaw too much, they become mushy and are likely to discolor.

Fruits that are to be cooked are handled like vegetables. In other words, cook them while they are still frozen.

How to freeze meats. All meats, including innards, smoked cuts, and game, freeze beautifully. There is really nothing to it. Just make sure the meat is of good quality; has been chilled below 40° within 24 hours of freezing; and is properly aged. You must also guard against:

1. Salmonellosis, a disease that is easily spread from food to man and back again. Once contracted, it causes severe illness and some-

times death. Foods most commonly contaminated include meats, poultry, fish, shellfish, eggs, and milk.

2. Staphylococcus. This is the dread infection that hospitals worry about. It most commonly contaminates ham, processed meats, poultry, beans, and milk products. The bacteria are present everywhere but usually get their start from nose, throat, and skin infections.

The best defense against both enemies is good sanitation at all times when putting up foods not only to be frozen but also by curing and smoking.

Keep your hands clean.

Never work around food when you are sick or have an infection.

Keep your hands away from your mouth, nose, and hair.

Cover coughs and sneezes with tissues.

Don't lick your fingers while working with foods and don't use cooking utensils to taste foods while preparing them.

Wash raw foods thoroughly before preparing them.

Wash utensils used for handling raw foods before you use them for cooked foods.

Keep all utensils and counters scrubbed. Wood cutting boards must be scrubbed especially hard and should be disinfected frequently to destroy the bacteria that lodge in cuts.

Since the bacteria multiply most rapidly at temperatures between 60° and 120°, one of the best ways to prevent bacterial growth is to keep cold foods cold and hot foods hot.

Having taken pains to control salmonellosis and staphylococcus, cut meats into sizes that fit your needs. Wrap them tightly in aluminum foil or drop them into polyethylene bags and squeeze out the air before sealing.

When packaging small cuts of meat, such as chops and hamburger patties, separate each from the other with sheets of foil.

Although meats can be cooked when frozen hard, you get better results by thawing them completely on the kitchen counter or, for fast results, in front of a fan. Don't remove the wrappings, but put the packages in pans to catch any juices that leak out. And if you have a dog, don't forget to push the pans well back on the counter out of his reach.

Allow the same cooking time as for fresh meat.

How to freeze poultry and game birds. It is generally best to dry-pick poultry and game birds. The process is greatly simplified for poultry if you kill the birds and simultaneously relax the feathers by tying their feet together; hanging them head down; inserting a thin, sharp, stilletolike knife through the mouth and into the brain; and giving the knife a quarter turn. This is easily done if the knife blade is inserted close to and parallel with the ridge in the top of the mouth.

To pluck the feathers, grasp small bunches firmly and pull against the grain. Take care not to tear the skin. Rub your thumb across the skin to remove the down. Pick out pin feathers between your thumb and a knife blade. Singe off any down and hairs that remain.

Another way to remove pin feathers is to paint melted paraffin on the birds with a clean paint brush. Then scrape off the wax gently with a paring knife held at an angle that won't cut the skin.

If you wet-pick poultry, kill it with a knife as above or chop off the head with an ax. Immerse the body in very warm (but not over 150°) water for about 30 seconds. Then pluck and singe.

To draw a bird, make a slit from the vent to the breastbone and pull out the entrails with your fingers. Don't overlook the lungs. Then make a cut near the base of the neck and remove the craw. Cut off the feet at the first joint and cut off the head and neck. Wash the body in cold water and drain thoroughly.

If the bird has been freshly killed and is still warm, put it into your refrigerator for several hours until it is completely chilled.

Wrap giblets and neck in a small polyethylene bag, piece of foil, or other moisture-vaporproof wrapping, and stuff them into the body cavity. Or if you prefer, you can freeze them separately. Then wrap the entire bird in aluminum foil or put it in a polyethylene bag.

A good trick when using bags for whole birds, as well as meats and fish, is to drop in the bird and dip the bag to three-quarters of its depth in hot—not boiling—water. This collapses the plastic onto the bird's body, thus expelling the air.

Although it is possible to stuff poultry before freezing, it is a poor practice because the stuffing and meat thaw at different rates and there is danger that bacteria may become active and poison you. Furthermore, any salt in the stuffing reduces the storage period drastically.

Before cooking frozen poultry, thaw it completely. Allow the same cooking time as for fresh poultry.

How to freeze fish. All fish freeze well; but the fresher they are, the better they will keep—and taste.

As soon as fish are out of the water, wash them thoroughly to remove the slime in which spoilage bacteria grow at a rapid pace. Then remove the scales, if any, and wash again.

Gut the fish, cut out the gills, and wash the body cavity, using 1 tablespoon chlorine bleach to 4 gallons water.

If possible, the fish should now be chilled as rapidly as possible in a refrigerator. If you're away from home, the best substitute for chilling is to rub a mixture of 1 tablespoon black pepper and 1 cup salt into the belly cavity at the rate of approximately 1 tablespoon per 3/4 pounds of fish. Then sprinkle the salt mixture on the skin, and place the fish in your creel or a box on a pad of damp seaweed or burlap. Cover the creel with more damp burlap (but don't let it touch the fish), and put it in a cool, shady place.

Small fish are usually frozen whole, with or without the head. Larger fish are filleted. Extremely large fish may be cut into steaks. Regardless of the treatment, dip the fish for a minute in a brine made of 2 tablespoons salt and 1 quart water; then wrap them tightly in aluminum foil or polyethylene bags after you have squeezed out the air.

A better method of handling cut-up fish after momentary brining is to lay them in pans and put them directly into the freezer without wrapping. (If the fish are fresh, no odors will be transferred to other frozen foods.) After they are frozen solid, dip them briefly in a solution made of 2 tablespoons white corn sirup and 1 quart water. Then return them to the freezer in the pans at once to solidify the dip. This is called glazing.

Repeat the process one more time. Then wrap the fish in foil.

A simpler way to glaze fish—though it takes up somewhat more freezer space—is to place the fresh fish in shallow aluminum freezer pans, cover them with water, freeze, and then wrap in foil.

Before cooking frozen fish, thaw them, preferably in the refrigerator. Cook them for the same time as fresh fish.

For directions on how to freeze shellfish, see the entry for the specific shellfish in the encyclopedic section.

15

How to Can Foods

The memory of that wondrous array will last forever. It wasn't a particularly bright basement. But the quart and pint jars which jostled one another on the shelves lining the north wall filled it with color and a solid sense of satisfaction. There were several kinds of beans, beets, squash, corn, Swiss chard, tomatoes, peaches, pears, applesauce, pickles, jams, and jellies. All the produce except the fruits had come from our own garden.

To say that this was our response to the war effort would not be quite correct. Ever since we had arrived in Connecticut, we had had a vegetable garden, and because we were anything but well off, it seemed a good idea to put up some of the food we raised. But after Pearl Harbor we went all out. And happily so.

Canning was rather fun. The results were tasty, nourishing, and amply plentiful. And somewhat to our surprise—and certainly much to our relief—we didn't have a single bad jar.

Today our enthusiasm for canning has been superseded by an even greater enthusiasm for freezing. But not everyone concurs with this feeling. For example, our neighbor John Seckla, who has one of the largest and finest fruit and vegetable gardens we have ever seen, remains true to canning even though he's had a freezer for years. He and his wife put up over 500 jars of food every year.

Canning equipment. Foods are canned in two ways. Those that are acid—fruits, tomatoes, sauerkraut, and pickled vegetables—are processed by the water-bath method. This means simply that the jars of food are cooked in boiling water for a specified period of time. The combination of the acid and 212° heat is enough to destroy the bacteria which might otherwise cause the food to spoil.

However, most vegetables, meats, poultry, and fish contain relatively little acid; consequently, in order to process them safely in a reasonable length of time, it is necessary to cook them under pressure at a temperature of 240°. This is called the steam-pressure method of processing.

The major piece of equipment needed for water-bath processing is a kettle deep enough to permit you to cover the jars with 1 to 2 inches of water and still leave space for the water to boil freely. Any kettle or even an old wash boiler will do, but the most suitable piece of equipment is a canning kettle with a cover and a wire rack to hold the jars. Depending on the size, these hold seven to nine quart jars.

For processing nonacid foods, you need a steam-pressure canner. Not an ordinary pressure cooker, because most of these are small and lack an adequate pressure control. A pressure canner is big enough to hold at least seven quart jars in a rack, and it has a pressure gage to permit accurate control of the temperature.

If a pressure canner is deep enough, you can also use it for water-bath processing simply by leaving the petcock open so steam can escape.

Containers. Some people put up food in steel cans. These save space in storage and their initial cost is low. But you must have a sealer to put on the tops. And since the cans are not reusable, the low cost is illusory.

Mason jars made of glass are better because you can use them forever if you don't crack them or chip the edge of the threaded top. They are most commonly sealed with a two-piece cap consisting of a flat disk with a rubbery sealing compound around the edge and a threaded band that screws down over the disk to hold it secure.

So-called regular jars are available in half-pint, pint, 1 1/2-pint, quart and half-gallon sizes. All have a mouth measuring 2 3/8 inches in diameter; consequently the caps are interchangeable.

Wide-mouth Mason jars have a somewhat larger opening. They come only in quart and half-gallon sizes.

Before packing food in a canning jar, you must inspect the jar and cap to make sure they are in perfect condition and will give an airtight seal. Wash the jar thoroughly in hot, soapy water and rinse it well. Wash and rinse the caps also. If the maker's directions call for it, dip the flat disks of the caps in boiling water.

General canning procedure. All foods to be canned should be of top quality and in prime condition. To prevent food poisoning, wash fruits and vegetables thoroughly, even though they are to be pared. When preparing meat, poultry, and fish, scrub all equipment and rinse with boiling water. Scrub and disinfect work surfaces.

Some foods are packed raw; some hot; and some both ways. Raw pack means that the food is packed into jars raw—uncooked—and is then processed. Hot pack means that the food is precooked briefly before it is packed into jars and processed.

How tightly food should be packed in jars depends on the packing method and the food. Raw fruits, for instance, are usually packed tightly. Meat, on the other hand, is packed loosely.

After placing the food in the jars, cover it with sirup, water, juice, or broth. Then to make sure there are no air bubbles, run a rubber scraper or table knife around the inside of the jar between the food and the sides of the jar. Also to remove bubbles, cut through the center of solid packs of corn, greens, and mashed pumpkins several times. Then add more liquid if necessary to cover the food completely. If not covered, the food at the top of the jar may darken. The top of the liquid should, as a rule, be somewhat below the rim of the jar. This is called the head space, and it varies with the type of food and jar size.

Put the cap on each jar as it is filled. First wipe the jar rim with a paper towel. Make certain the sealing compound on the disk section of the cap is clean. Then place the disk on the rim and screw the metal band down over it until the band is tight. Since this kind of cap is a self-sealer, you do not have to tighten it further, even after processing.

Place the capped jars in the canning kettle rack, lower them into the kettle, and process according to the general directions that follow in this chapter and the specific directions given for each food in the encyclopedic section. If liquid is lost during processing, don't worry; and don't try to replace it.

After processing, remove the jars from the kettle and cool them, top side up, on a metal rack, folded cloth, pad of newspapers, or a wood-cutting board. Leave a little space between jars for the air to circulate, but keep them out of drafts.

When the jars are cool, test the seal by pressing down on the center of the cap. If it is already slightly concave or if it stays down when pressed, the seal is good. You should then remove the metal band, because if it is left in place, it may rust and be hard to loosen.

If a jar is not sealed tightly, put it in the refrigerator and eat the food as soon as possible. It is not yet spoiled but will become spoiled in a short while. The alternative is to empty the jar, wash it, pack it full of the food again and process a second time. This is a poor idea, however, because the food is overcooked by reprocessing.

Label and date jars, and store them in a cool, dry, dark place. The foods will keep almost indefinitely if the seal is not broken. After a year, however, quality begins to decline.

Before using canned foods, examine the cap and container to make sure they are sound. If the lid is bulging slightly, the food has probably spoiled. When you remove the lid, look for mold and sniff for an off odor. If you suspect food has spoiled, toss it out at once. Don't taste it.

Food containing the poison that causes botulism may not, unfortunately, look unusual. But you have no reason to worry about botulism if you know your pressure canner and gage were in perfect condition when you put up the food. If you are not certain about this, however, you should boil the food—at least 10 minutes for most vegetables; 20 minutes for corn, spinach, meat, poultry, and fish.

How to can fruits, tomatoes, sauerkraut, and pickled vegetables. Prepare foods as specified in the encyclopedic section and pack in jars.

Fruits can be canned in their own juice, extracted juice, or water —without sweetening. The process is the same as if they are sweetened. However, most people prefer sweetened fruit because it holds its flavor, color, and shape better.

To make sugar sirup, boil sugar in water or juice extracted from some of the fruit. Use proportions given in Chapter 14.

Use the sirup called for in the encyclopedic section, or suit your own taste. In all cases, the sirup should be hot.

If you are putting up very juicy fruit by the hot-pack method, add about 1/2 cup sugar directly to each quart of raw, prepared fruit, heat to simmering, and pack the fruit in the juice that cooks out.

If you prefer, you can substitute light corn sirup or mild honey for up to half of the sugar called for in the sirup formulas in Chapter 14.

To keep fruits such as apples and peaches from darkening during preparation, drop them after skinning into a solution of 1 gallon water, 2 tablespoons salt, and 2 tablespoons vinegar. Don't let them soak in this more than 20 minutes and be sure to rinse them in water before packing.

A more reliable antidarkening process and a better one since it also prevents discoloration of fruit in the jar is to mix a prepared ascorbic acid mixture, such as Fruit-Fresh, with your sirup according to the maker's directions. Pour some of this into the Mason jars and drop in the fruit as you prepare it.

After sealing the jars, place them in the canning kettle. If you are processing by the raw-pack method, the kettle should be partially filled with hot water; for the hot-pack method, it may contain boiling water. Add enough hot or boiling water to cover the jars an inch or two. Cover the kettle and turn up the heat. When the water comes to a rolling boil, start timing. Make sure the water boils steadily throughout the processing period. If necessary add boiling water to replace whatever boils out. When processing time is up, remove the jars from the canner at once.

Processing times given in the encyclopedic section are for people who live at an altitude of less than 1,000 feet. If you live at higher altitude, processing time should be increased.

If you live at an altitude of	Increase processing time if the time called for is:	
	20 minutes or less	More than 20 minutes
1,000–2,000 feet	1 minute	2 minutes
2,000–3,000	2	4
3,000–4,000	3	6
4,000–5,000	4	8
5,000–6,000	5	10
6,000–7,000	6	12
7,000–8,000	7	14
8,000–9,000	8	16
9,000–10,000	9	18
over 10,000	10	20

How to can low-acid vegetables. Prepare foods as called for in the encyclopedic section, pack in jars, and cover with the boiling water or liquid in which you precooked vegetables for hot packing. After sealing, place the jars in the rack in the pressure canner.

Follow the directions of the manufacturer for using the pressure canner. But the procedure in general is as follows:

Pour 2 to 3 inches of boiling water in the canner before lowering in the rack of filled jars. Make sure the jars are placed so that steam can circulate around them freely. If processing two layers of small jars, the upper layer should be in a rack and the jars should be staggered so as not to be directly over those in the bottom layer.

Tighten the cover but leave the vent open and allow steam to escape for 10 minutes or more. Then close the vent. When the pressure reaches 10 pounds, as shown on the gage, start timing. To maintain constant pressure, raise and lower the heat under the canner. Don't let drafts blow on the canner.

As soon as processing time is up, remove the canner from the range and let it cool until the pressure has dropped. Don't rush this process. When the pressure gage registers zero, wait a couple of minutes, and open the vent slowly. Then slowly remove the cover, holding it away from you so that you will not be exposed to the escaping steam. Finally, lift out the jars.

As noted above, the normal pressure for processing vegetables is 10 pounds. At this pressure, the temperature in the canner is 240°. However, you should process at this pressure only if you live at an altitude of 2,000 feet or less. At higher altitudes it is necessary to increase the pressure in order to reach 240°. The amount to increase is shown in the following table.

If you live at an altitude of	Process foods at a pressure of
2,000–4,000 feet	11 pounds
4,000–6,000	12
6,000–8,000	13
8,000–10,000	14
over 10,000	15

How to can meat and poultry. Freshly killed meat and poultry should be chilled thoroughly immediately after slaughter. Remove as much fat as possible, and cut the food into pieces that will fit in the Mason jars.

If meat and poultry are to be processed by the raw-pack method, fill the jars loosely to within 1 inch of the top. Place the food up and down rather than crosswise in order to allow for good circulation of the liquid. Add 1 teaspoon salt per quart jar (this may be omitted if you don't want a salty product, however). Place the open jars in a kettle filled with water to within 2 inches of the jar tops. Cover the kettle and boil the water slowly until the temperature at the center of the jars registers 170°. This takes about 50 to 75 minutes and drives the air out of the food. Then wipe the jar rims clean, put on caps and tighten them, and process the jars in a pressure canner at 10 pounds pressure (or more if you live at a high altitude). Follow the directions for processing vegetables.

If meat and poultry are processed by the hot-pack method, make a broth from the bones and scrap pieces, and skim off the fat. Cook the food which is to be canned in the broth until it reaches 170° or a little higher. Then immediately pack the food loosely in jars, add salt, and pour the boiling broth over the meat to within 1 inch of the top. Work out the air bubbles. Clean the jar rims, apply caps, and process the jars at 10 pounds pressure according to the directions for processing vegetables. Processing times are given in the encyclopedic section.

Before eating any home-canned meat or poultry, boil it for 20 minutes.

How to can seafood. The process varies slightly, depending on the seafood. In all cases, however, the food must be processed in a pressure canner at 10 pounds pressure.

For specific directions, see the encyclopedic section.

16
How to Make Wine

There is a scientific way to make wine and a simple way. Since the scientific way produces superior wine, we should describe that process here. But, unfortunately, there isn't space: a whole book is needed to do the subject justice. So we'll stick to the simple way.

If it doesn't produce wine that would command $8 a bottle, it does nevertheless produce a good to better-than-average wine; and since people have been making and enjoying that kind of wine for centuries, there is no reason why you shouldn't, too. In fact, we predict you will.

Government approval. Federal law permits the "duly registered head of any family" to make up to 200 gallons of wine annually for family use without payment of tax. Before you start making wine, however, you must secure the government's authorization.

This is easily done. Just write to the Alcohol and Tobacco Tax Division of the Internal Revenue Service in the office nearest you and ask for two copies of Form 1541. After filling these out, return them to the division for approval. As soon as one of the copies is mailed back to you, you can go to work.

While waiting for this authorization to arrive, you should also find out what, if anything, your state has to say about wine making. If you can't ferret this information out of the state statute books at your local library, write to your state tax department.

Equipment needed. The principal pieces of equipment needed for wine making are a large, open container for the initial fermentation of the fruit you are using; another large, closed container for the second fermentation; and a similar, large, closed container for "racking" the wine. These containers are often called "fermenters."

As a rule, home wine makers use an earthenware crock as their primary fermenter. But the problem with eathenware—and also with glass—is that it is heavy and breakable. For that reason, you might be smart to use an enameled canning kettle provided the enamel is not chipped or cracked. An even better container is one made specifically for the purpose out of heavy-duty polyethylene.

The secondary fermenter is normally a carboy—a very large, glass bottle similar to those in water fountains. Here again, however, you can get a polyethylene jug which weighs much less. An excellent alternative is to use 1-gallon glass jugs, which are easy to come by and easy to handle. They also permit you to make wine in varying quantities rather than in a batch large enough to fill a carboy.

Other essential equipment includes the following:

A fermentation lock. This is a small device which is placed in the opening of the secondary fermenter to allow gas to escape while preventing air from entering. You can buy one ready-made from a wine maker's supply outlet. Or you can make your own out of a big cork with a small glass tube running through from top to bottom. At the top, the tube is connected to a small, short rubber hose which is bent down and stuck into a glass of water.

A small-diameter hose for siphoning wine from one container to another.

A supply of green or amber bottles with long, straight necks which provide good contact with the corks.

All bottles, corks, and other equipment must be spotlessly clean; otherwise the wine will pick up off flavors. You should not, however, use soap or detergents for cleaning because, if they are not thoroughly rinsed away, they will contaminate wine, too. A hot, strong solution of washing soda is best. Scrub it on and let it stand awhile if things are very soiled. Then rinse your equipment in hot water twice and let it drain upside down. Don't dry wth a towel, because it may be dirty.

Crushing the fruit. Although we speak from here on only about making grape wine, we do not overlook the fact that many, many other fruits as well as flowers and vegetables, are made into wine. We concentrate on grapes only because it is simpler to talk about one fruit than a long list. Actually, all wines are made in roughly the same way. Whatever differences there are will be found in the directions in the encyclopedic section, where we describe the manufacture of a few of the more common wines.

Select grapes that are fully ripe. Don't wash them unless they are obviously dirty or have recently been sprayed with insecticide or fungicide. The white bloom on the berries should be preserved as much as possible because it contains the yeasts that stimulate fermentation. You should, however, pick over the berries and discard any that are unripe, spoiled, or split. As you do this, strip the berries from their stems; but if you miss some of the stems, don't worry. Some people say grape wine is best made with the stems on the fruit; some say it is best made with the stems off.

If you own a press, use this to crush the grapes, but don't exert so much pressure that you also crush the seeds. If you don't have a press (until you get to be a wine-making expert, it won't make much difference whether you have a press or not), simply pour the grapes into your open fermenter and crush them thoroughly, layer by layer, with a potato masher. Fill the fermenter no more than three-quarters full.

Fermenting the fruit. At this point, many wine makers add white sugar to the crushed grapes—which are called "must"—in order to raise the sugar content of the juice so that it will produce sufficient alcohol. This is usually necessary with other fruits made into wine; but grapes as a rule contain enough natural sugar to ferment actively without assistance.

Another thing often added at this point is dried wine yeast—not baker's yeast. But here again, unless grapes are strangely lacking in bloom, it is usually unnecessary to augment the supply of yeast.

In other words, after crushing the grapes, let them ferment by themselves. You should, however, cover them with cheesecloth to keep out bugs. And keep an eye on the temperature of the room. Ideally it should be about 70°. Don't let it go over 80° or under 50°.

The must will start to ferment in 24 hours, as a rule; but it may start earlier and it may be considerably slower. If it is very slow to start, the chances are the room temperature is too low.

In any event, once fermentation gets under way, it becomes quite violent; and the grape skins will come popping to the surface, where they form a floating cake, or cap. This should be pushed down into the juice at least once and preferably twice a day.

After a number of days, fermentation will subside and you can skim off the cake. If you have a press, put the cake in it, squeeze out whatever juice it contains, and return this to the fermenter. Tasting the juice in the fermenter, you will find it to be quite sour; so you should now add sugar—just enough to give the juice a pleasant but not strong sweetness.

With the addition of the sugar, fermentation will become vigorous again and will continue so for several days. When it subsides, stick your siphon hose down into the juice to a point just above the sediment, or "lees"; put your secondary fermenter on the floor and cover the mouth with cheesecloth; then siphon the juice into it. Add a little more sugar if the juice is sour. For a good dry wine, the juice should be a little sweet.

Fill the secondary fermenter to about 90 percent of capacity, and put any juice that is left over into your refrigerator. Then close the top of the fermenter with the fermentation lock and let the fermentation visible in the container die down completely. This should take only a few days. You must then open the fermenter and fill it all the way to the top. To do this, use the left-over juice from the refrigerator. If you don't have any left-over juice, drop washed pebbles into the fermenter to raise the liquid level.

From here on, until the wine is transferred into small bottles for serving, the large containers in which it is stored must be kept full. If they are not, the wine will spoil.

Racking and fining the wine. Store the secondary fermenter in a cool place for about 3 months, or until bubbles stop forming in the fermentation lock. You should then siphon the wine into your other large, closed container. This is called racking. Take pains not to transfer the sediment from the bottom of the fermenter along with the clear wine. Fill the new container to the top by dropping in pebbles or adding a similar wine. Then close the container with the fermentation lock and store it in a cool place for another couple of months. As soon as bubbling (if any) stops in the fermentation lock, replace it with a solid cork.

The wine can now be drunk, but it will not be as smooth as you might like, and it will be rather cloudy. There are two ways to get rid of the cloudiness. If you're not impatient to sample your wine, use them both.

The first way is to rack the wine two or three more times at 2-month intervals.

The other way is to fine the wine. This consists simply of adding to the wine a chemical that removes the sediment by precipitation. Various materials such as charcoal, isinglass, and egg white are used, but ordinary unflavored gelatin is about as good as any. Use 1 ounce per 5 gallons wine. Dissolve it in a little warm water, pour it into the wine, and let it work for a week or two. Then siphon off the wine into another container.

Bottling. When the wine is clear, siphon it into clean, sterilized, well-drained bottles. Put the outlet end of the siphon hose in the bottom of the bottles so that the wine will not pick up any more oxygen than necessary during the transfer process. Fill the bottles to within 3/4 inch of the bottom of the cork.

Before corking, soak the corks in warm water until they are soft, then dry them. If you use cork stoppers with caps, they can be put in by hand; but if you use long, cylindrical corks, you will need to buy a simple corking device to drive them down flush with the bottle rim.

Label the bottles and store them in a cool place on their sides so that the corks will not dry out and let oxygen seep in. As a rule, red wines are best if stored in bottles for a year or two. White wines, on the other hand, do not improve greatly in glass.

Encyclopedia
of Foods Grown and Preserved in the United States

ABALONE

Abalone is a handsome, large mollusk prized along the Pacific Coast for its richly delicious meat and everywhere else for its shell, which is lined with mother of pearl. After prying it loose from its salt-water feeding grounds, cut the big steak-like bottom muscle, or foot, from the shell; trim off dark areas and sole of foot; rinse, and wrap in a clean cloth. Lay the meat flat on a solid surface and pound it—but not too hard—with a mallet five or six times till it is limp. Then brine, glaze and freeze like fish, as in Chapter 14. Store for 6–9 mo.

ACEROLA

Also known as Barbados cherry, acerola is a smallish evergreen shrub with bright red fruits like cherries. These have a higher Vitamin C content than the fruits of any other plant.

JELLY

Makes 7–8 8-oz. glasses

2 lb. acerolas
4 cups water
7 cups sugar
1 bottle liquid pectin

Wash fruit but don't stem. Measure 4 cups into a kettle, crush, and add water. Bring to a boil, then simmer till fruit is soft —about 15 min. Extract juice. Measure 2 1/2 cups into a kettle and mix with sugar. Follow standard procedure for making jelly with liquid pectin.

BUTTER

Makes 4 8-oz. glasses

3 cups acerola pulp
3 cups sugar
3 tsp. finely chopped fresh ginger

Use pulp left from making jelly. Put it through a food mill and measure 3 cups into kettle. Mix in sugar and ginger. Simmer, stirring constantly, until thick. Then pour in hot, sterilized glasses and seal.

ALBACORE

Albacores are large mackerels and are sometimes known as white-meat tuna. Freeze and can like tuna.

ALEWIVES

Alewives are herrings that spawn in rivers and then return to the sea, where they reach a weight of about 1 pound. They are best processed by curing, smoking, and canning. Follow directions in Chapter 4. Cold smoking is preferable to hot.

Alewives may also be salted by the standard processes and pickled like herring.

ALMONDS

The nuts on almond trees mature from the outside of the tree in. The best time to collect them is when almost all the inside nuts are mature. Maturity is indicated by splitting of the hulls. The shells and kernels then start to dry out. Many of the nuts fall from the trees. The rest can be picked or knocked down with long bamboo poles.

Let the nuts dry under the trees for 2 or 3 days. Then remove the hulls and spread the nuts out in an airy, shaded place to dry several days more. After drying, store them in sacks in a cool, dry place.

If kernels are to be stored, blanch them by covering with boiling water for 2 min. Then remove to cold water, rub off skins, and dry the kernels on paper towels. The kernels are then ready for freezing or canning, according to directions in Chapter 8.

ANCHOVIES

Anchovies are small fish of the herring family found in both the Atlantic and Pacific oceans. They can be cured in salt and stored dry or stored in salt brine. But they are best when cured, smoked, and canned.

For canning without smoking, see *Sardines.*

ANISE

Anise is an annual herb with a mild licorice flavor. You can dry the leaves and use them to flavor other foods, but the seeds are the important product. Collect them when dry and store in a tight bottle. Use in candies and cookies.

ANTELOPE

Freeze or smoke antelope or make it into jerky. Process like venison.

APPLES

If apples are to be dry-stored, they should be picked just before they are fully ripe. Keep them in a cool, somewhat humid, dark root cellar. They can also be held for several months in the fresh-food compartment of a refrigerator, but unless you own a couple of refrigerators, it's hardly likely you have space for the fruit.

Some of the varieties that keep especially well in dry storage are Baldwin, Cortland, Monroe, Northern Spy, Rhode Island Greening, Rome Beauty, Spartan, Spigold, Stayman, and Winesap. Idared is outstanding.

FREEZING

Sugar pack. Wash, peel, core, and slice firm apples. To prevent darkening, drop slices into 1 gal. cold water containing 2 tbs. salt and 2 tbs. vinegar, but don't hold them in this solution for more than 15 min. Drain. Blanch for 2 min. in boiling water. Chill thoroughly. Mix 5 cups apple slices with 1/2 cup sugar and 1/2 tsp. ascorbic acid. Pack and freeze. Use for pies. Store for 10–12 mo.

Dry pack. Follow directions for sugar pack but omit sugar. Use for pies. Store for 10–12 mo.

Sirup pack. Wash, peel, and core apples and slice them directly into rigid containers partially filled with 40 percent sugar sirup to which ascorbic acid has been added at the rate of 1 tsp. per cup. Cover fruit with sirup, seal, and freeze. Use for fruit cup. Store for 10–12 mo.

Applesauce. Wash fruit and cut into quarters. Cook with a very small amount of water till soft. Press through a food mill. Let cool thoroughly. Add sugar and cinnamon to taste. Pour into rigid containers to 1/2 in. of top, and freeze. Store for 10–12 mo.

CANNING

Wash, peel, core, and slice apples into 1 gal. water containing 2 tbs. salt and 2 tbs. vinegar. Let stand no more than 15 min. Drain. Boil for 5 min. in water or a 30 percent sugar sirup. Pack in jars to 1/2 in. of top and cover with sirup to 1/2 in. of top. Seal. Process in boiling water. Pints for 15 min.; quarts for 20 min.

Applesauce. Wash and cut apples in quarters. Cook with very little water until soft. Press through a food mill and season to taste. Reheat to simmering, stirring constantly, and pack in jars to 1/4 in. of top. Seal. Process in boiling water for 10 min.

DRYING

Wash, peel, core, and slice apples thin. Drop slices into a solution of 1 gal. water, 2 tbs. salt, and 2 tbs. vinegar. Sulfur for 15–25 min. Place in trays no more than two layers deep and dry in the oven until springy and glovelike. Start at 130°, increase to 165°, and finish at 145°. Place in tight containers and store in a cool, dark place.

JELLY

Without added pectin
Makes 3–4 8-oz. glasses

3 lb. apples
3 cups water
3 cups sugar

Wash apples; remove stem and blossom ends; cut into small pieces, and place in kettle with water. Cover, bring to a boil, then simmer until apples are soft, about 20 min. Extract juice. Measure 4 cups into kettle and mix in sugar. Then follow standard procedure for making jelly without added pectin.

JELLY

With added pectin
Makes 11 8-oz. glasses

4 lb. apples
6 1/2 cups water
7 1/2 cups sugar
1/2 bottle liquid pectin

Wash; remove stem and blossom ends; cut into small pieces; place in kettle with water; bring to a boil, then simmer until fruit is soft. Extract juice. Measure 5 cups juice into kettle, mix with sugar, and follow standard procedure for making jelly with liquid pectin.

APPLE BUTTER

Without added pectin
Makes 5 pints

6 lb. apples
2 qt. sweet cider
3 cups sugar
1 1/2 tsp. ground cinnamon
1/2 tsp. ground cloves

Wash fruit; remove stem and blossom ends; cut into small pieces, and cook in cider till soft, about 15 min. Put through a food mill. Measure 3 qt. pulp into a kettle and cook slowly, stirring frequently, until about half as thick as desired. Add sugar and spices and continue slow cooking and stirring until liquid does not separate from the butter when a spoonful of the latter is put on a cold plate. Then ladle into clean, hot Mason jars to 1/2 in. of top. Seal. Process in boiling water for 10 min.

APPLE BUTTER

With added pectin
Makes 11 8-oz. glasses

5 cups apple pulp
7 1/2 cups sugar
1 tsp. ground cinnamon
1/2 tsp. ground allspice
1/2 bottle liquid pectin

When apple butter is made with commercial pectin, there is no need to boil down the mixture to make it thick; consequently, it is best to start with pulp from which juice has been extracted.

Extract juice from 4 lb. apples as in recipe for making jelly with added pectin. Save juice for jelly. Press pulp left in jelly bag through a food mill or sieve. Measure 5 cups into a large kettle and add spices and sugar. Mix well, place over high heat, and follow standard procedure for making jam with liquid pectin.

CHUTNEY

Makes 8 8-oz. glasses

12 medium, tart apples
3 large green peppers
6 green tomatoes

4 small white onions
1 cup seeded raisins
2 tbs. mustard seed
2 cups sugar
1 qt. vinegar
2 tsp. salt

Wash, core, peel, and coarsely chop apples. Remove seeds. Core peppers and chop. Chop tomatoes and onions. Mix all ingredients together in a kettle and cook slowly until of good consistency—about 30 min. or more. Pack in hot, sterilized jars and seal.

PECTIN

7 large, tart apples
4 cups water
2 tbs. lemon juice

Wash and stem apples and cut into pieces. Boil with the water and lemon juice till soft, then press through a jelly bag and strain resultant juice through a clean flannel bag without squeezing. Boil for 15 min. Pour into hot, sterilized jars and seal. Store in refrigerator or a very cool, dark room. In making jelly, use 1 cup apple pectin with each cup of the other fruit juice.

CIDER

Not everyone has the same understanding of what cider is. Some think it contains alcohol; some think it does not. To clarify matters (at least for the purposes of this book; we can't hope really to settle the argument), we define cider as sweet, nonalcoholic, processed apple juice. Apple juice which has been fermented and contains alcohol is hard cider, a wine.

If you want to make cider, you can take your apples to a cider mill and have them pressed and bottled for you. Or you can buy a small antique cider press or a new 8-qt. model from Sears, Roebuck. We'll assume you make this entirely a do-it-yourself project.

Apples used for cider need not be beautiful, but they should be sound and ripe enough to eat out of hand. If you use culls —either from the tree or windfalls—cut out the bad spots.

Late apples are the best for cider making, though you need not stick to them 100 percent if you have a tasty, early variety and allow it to ripen completely. In any case, use a blend of three or four varieties to get the tastiest cider.

After washing and picking over apples, put them through the cider mill. The first step is to grate or chop the fruit into a coarse pulp. The second step involves pressing the juice out of this pulp, which is called "cheese." The way you carry out these steps depends on the design of your cider mill.

When the juice has been pressed out and dripped into a clean container, strain it through a jelly bag or four or five layers of washed cheesecloth into another container. Let it settle for about 24 hr., or until it looks clear. Then siphon off the clear juice without disturbing the sediment in the bottom of the container. The juice is now ready for packaging.

For short-term storage, simply pour the juice into sterilized bottles and store them in a refrigerator. They will keep for up to 2 wks.

A much better storage method is to pour the juice into clean bottles or rigid plastic containers. Fill only nine-tenths of the way. Then seal and store in the freezer. The juice will keep without losing its fresh-from-the-mill flavor for a year.

A third way to handle the juice is to put it in the top of a double boiler; place over boiling water; and heat the juice, stirring often, to 190°. Use a thermometer. Then bottle according to directions in Chapter 9.

HARD CIDER

The most delightful description of hard cider we have even run across was written by J. M. Trowbridge in *The Cider Makers' Hand Book* (published in 1897). It reads in part as follows:

> A pure article of cider should have perfect limpidity and brightness; it may vary in color from a delicate straw to a rich amber, but should never show much of a roseate tinge. It should be fragrant, so that when a bottle is freshly opened and poured an agreeable, fruit perfume will arise and diffuse itself through the apartment "with a benison on the giver". It should be tart, like Rhine wine, and by no means sharp or harsh. It should have a pleasant, fruity flavor, with aromatic and vinous blending, as if the fruit had been packed in flowers and spices. It should have a mild pungency, and feel warming and grateful to the stomach, the glow diffusing gradually and agreeably throughout the whole system, and communicating itself to the spirits. It should have a light body or substance about like milk, with the same softness and smoothness, and it should leave in the mouth an abiding agreeable flavor of some considerable duration, as of rare fruits and flowers.

To make this beautiful drink, select and blend apples with great care and put them through your cider mill. To remove all undesired particles, toss a box of white, unscented facial tissues into 2 qt. boiling water; let stand for a minute; then beat them into small pieces with a fork. Pour the pulp into a strainer and press it as dry as possible. Then place it in a clean flannel bag and slowly pour the apple juice through it into a large, open fermenter, or vat. Let it stand.

Within a day or two, the juice will start to work. When it forms a tenacious, frothy scum on the surface, start skimming every couple of hours. Continue this until the scum no longer comes to the surface. At this point the juice is in active fermentation and should be siphoned into a clean, hardwood barrel or plastic fermenter with a small opening. The barrel or fermenter should just previously have been rinsed with hot water to raise its temperature a bit. Don't let the juice chill during the transfer into the barrel. Fill the barrel 80 to 90 percent and put the bung in *loosely.*

Store the barrel in a room at a temperatue of 70°. Hold the temperature as steady as possible. Inspect the cider every 8 hr. When the fermentation finally stops, siphon it off through a tube running through a pan of ice and a facial tissue filter into a chilled barrel. Drive the bung in tight and store in a cool place for a couple of months before using.

VINEGAR

The easiest way to make vinegar is to start out as if you were making hard cider, and let the cider continue to ferment in an open container until it turns sour. A better procedure is to follow the directions for making hard cider up to the point where the cider is siphoned from the first fermenter. Leave the bung hole in the second fermenter open (but cover it with mesh to keep out insects) until the cider turns sour.

Once you have made a first batch of vinegar, you can set some of it aside—before pasteurization—to hasten production of a second batch. In this case, after you have transferred the cider from the first to the second fermenter, pour in the unpasteurized vinegar at the rate of 1 part vinegar to 5 parts cider. Leave the top of the fermenter open but screened. Store the fermenter in a room at 75° or a little higher.

Once the cider turns to vinegar, the vinegar can be used at once; but it is improved by letting it age for two weeks or more. This is accomplished by filling the fermenter to the top and capping it tightly.

To store vinegar—whether newly made or aged—for any length of time, heat it to 150° in a kettle; then pour it immediately into hot, sterilized bottles, and seal.

APRICOTS

FREEZING

Sirup pack. Pack apricots that are to be eaten uncooked in this way. Wash, peel, halve, and pit the fruit. Pack with 40 percent sugar sirup to which ascorbic acid has been added. (Use 1 tsp. ascorbic acid per 1 cup sirup.) Store for 10–12 mo.

Sugar pack. Pack this way if apricots are to be cooked. Wash, peel, halve, and pit fruit. Drain and mix 4 to 5 lb. fruit with 2 cups sugar to which 2 tsp. ascorbic acid have been added. Store for 10–12 mo.

Puréed apricots. Wash, pit, quarter, and press fruit through a food mill or purée in a blender. Pack unsweetened, or mix 4 to 5 lb. purée with 1 lb. sugar. Adding 3 tsp. ascorbic acid is advisable. Store for 10–12 mo.

CANNING

Raw pack. Fruit can be left unpeeled if you wish. Cut in halves and pit. To prevent darkening, drop into solution of 1 gal. water, 2 tbs. salt, and 2 tbs. vinegar. Pack in jars to 1/2 in. of top. Cover with boiling, 40 percent sugar sirup to 1/2 in. of top. Seal. Process in boiling water. Pints for 25 min.; quarts for 30 min.

Hot pack. Prepare as above. Heat apricot halves through in 40 percent sugar sirup. Pack fruit in jars and cover with sirup to 1/2 in. of top. Seal. Process in boiling water. Pints for 20 min.; quarts for 25 min.

DRYING

Use fruit which is fully ripe but not soft. Wash, halve, and pit. Sulfur for 3 hr. Or if you prefer a "candied" product, simmer for 10 min. in sirup made of equal parts sugar and water. Then remove from fire and let stand in sirup for 10 min. before draining. Spread treated slices pit side up in single layers in trays, and heat in oven for about 6 hr. until leathery. Start at 130° and gradually increase to 150°. Keep oven door shut as much as possible in order to prevent too much moisture from escaping.

JUICE

Use firm fruit. Wash and stem, and drop into 1-in.-deep boiling water. Cook till soft. Press pulp through a sieve and then strain juice through cheesecloth. Mix with equal parts of sirup made with 1 cup sugar to 4 cups water. Then follow standard juice-making procedure.

JAM

Makes 8 8-oz. glasses

2 1/2 lb. apricots
1/3 cup lemon juice
6 1/2 cups sugar
1/2 bottle liquid pectin

Wash and pit fruit; cut in small pieces and grind. Measure 3 1/2 cups into kettle. Mix in lemon juice and sugar. Follow standard procedure for making jam with liquid pectin.

ICE CREAM

See *Milk.*

CANDIED APRICOTS

See Chapter 10. Peel apricots, cut in half, and remove pits.

ARTICHOKES

Some things are a lot better fresh than preserved, and artichokes are one of them. But this doesn't mean you should not put them up.

FREEZING

If freezing the hearts only, remove all leaves and trim stems. If freezing small, whole artichokes (under 2 in. in diameter), remove outer leaves, cut off bristly tops, and trim stems. Blanch for 4 min. Follow standard freezing procedure. Store for 10–12 mo. To serve, boil for 10–12 min.

PICKLED ARTICHOKES IN OIL

Makes 4 pints

4 doz. artichokes about 3 in. in diameter or
6 doz. large artichokes
1 cup lemon juice
2 qt. water
cider vinegar
4 cloves garlic
8 bay leaves
1 tsp. basil
1 tsp. oregano
4 cups olive oil

Remove all leaves from large artichokes, leaving just the hearts. Remove tough outer leaves of small artichokes and cut off the bristly tops. Trim stems. Wash thoroughly. Place in a kettle containing lemon juice and water. Boil small, whole artichokes 10 min.; hearts of large artichokes, 5 min. Drain and pack in clean pint jars. Cover with vinegar and let stand for 12 hr. Drain, cover with fresh vinegar, and let stand 4 hr. more. Drain. Add 1 clove garlic, 2 bay leaves, 1/2 tsp. basil and 1/2 tsp. oregano to each jar and fill to within 1/2 in. of top with oil. Seal. Process in boiling water bath for 30 min.

PICKLED ARTICHOKES IN WINE VINEGAR

Makes 4 pints

4 doz. artichokes about 3 in. in diameter or
6 doz. large artichokes
1 cup lemon juice
2 qt. water
cider vinegar
4 cloves garlic
8 bay leaves
1 tsp. basil
1 tsp. oregano
1 qt. wine vinegar

Remove all leaves from large artichokes, leaving just the hearts. Remove tough outer leaves of small artichokes and cut off bristly tops. Trim stems. Wash. Place in kettle with lemon juice and water. Boil small artichokes for 10 min.; hearts of large artichokes for 5 min. Drain and place in clean pint jars. Cover with cider vinegar and let stand for 12 hr. Drain. Divide spices equally among jars and fill to within 1/2 in. of top with wine vinegar. Seal. Process in boiling water bath for 20 min.

ASPARAGUS

FREEZING

Sort asparagus into large and small sizes. Peel the bottom half or two-thirds of large stalks. If packing in rigid containers, cut the stalks 1/2 in. shorter than the height of the containers. Longer stalks can be wrapped in aluminum foil. Stalks can also be cut into 2-in. lengths and packaged in rigid containers. Follow standard freezing procedure. Blanch for 3 min. (4 min. for very large stalks). Alternate tips and stem ends when packing; or if using rigid containers which are wider at the top than at the bottom, pack tips down. Store for 10–12 mo. To serve, cook for 5–8 min.

CANNING

Raw pack. Sort according to size. Peel the bottom half or two-thirds of large stalks. Cut in 1-in. pieces. Pack tightly into jars, but don't crush. Add 1/2 tsp. salt to pints; 1 tsp. to quarts. Cover with boiling water to 1/2 in. of top. Seal. Process in pressure canner at 10 lb. pressure. Pints for 25 min.; quarts for 30 min.

Hot pack. Prepare as above. Place in boiling water and boil for 3 min. Pack loosely into jars. Add 1/2 tsp. salt to pints; 1 tsp. to quarts. Cover with cooking water to 1/2 in. of top. Seal. Process in pressure canner at 10 lb. pressure. Pints for 25 min.; quarts for 30 min.

ASPARAGUS PEAS

This vegetable is a small annual plant which produces odd, stringy pods that are harvested and eaten whole when 2 1/2 to 3 in. long. Freeze or can like string beans.

AVOCADOS

Soft, ripe avocados that are free of blemishes can be frozen as a purée. Simply peel and pit the fruit, mash the pulp, and measure. Mix 4 cups pulp with 1 tsp. ascorbic acid, and pack in rigid containers. Use for salads and sandwiches. For a sweet purée to be used in ice cream, mix 4 cups purée with 1 cup sugar to which 1 tsp. ascorbic acid has been added. Store for 10–12 mo.

BANANAS

FREEZING

Remove skins and strings from ripe, unblemished fruits. Mash. Make a sirup in the proportions of 1 cup sugar to 4 cups water, and to this add 4 tsp. ascorbic acid. Mix this thoroughly with the fruits at the rate of 1 to 2 tbs. per banana. Package in rigid containers, seal, and freeze. Store for 4–6 mo.

ICE CREAM

See *Milk.*

BANANA BUTTER

Makes 8 8-oz. glasses

10 bananas
2 tbs. lemon juice
6 1/2 cups sugar
1 bottle liquid pectin

Mash peeled, stringed bananas and measure 3 cups into a kettle. Add lemon juice and sugar. If you wish, you can also add about 1/4 cup finely chopped maraschino cherries. Mix thoroughly and bring to a rolling boil. Boil hard for 1 min., stirring constantly. Remove from heat and stir in pectin at once. Ladle into hot, sterilized glasses and seal.

BARLEY

Barley resembles wheat and has the same sort of long bristles coming up out of the seed spikes. The grain lacks the gluten which is needed to make raised bread, but it is ground into flour and made into black bread—especially by Europeans. Thresh, winnow, and mill as in Chapter 7.

Barley grain is most commonly made into malt for brewing, but, unfortunately, that's a job that requires more equipment and patience than the average individual has.

BARRACUDA

This popular game fish is also a good food fish. Wrap fillets in aluminum foil, freeze, and store for 4–6 mo.

The fish is also split lengthwise to, but not through, the back skin and folded out flat. Make deep slits in the flesh, and dredge with salt. Cure in a large container for 48 hr. Then wash and dry the fish outdoors as described in Chapter 5. When you bring the fish in at night, place a thick pad of clean sacking between each layer, and weight the pile down heavily. Continue drying and stacking in this way for about 5 days, until the fish is dry. Then wrap in wax paper, pack in tight boxes, and store in a cool, dry place.

BASIL

Basil is an annual herb which is easily and quickly grown almost everywhere. The best variety for use in flavoring soups, salads, stews, meats, and egg and tomato dishes is a green-leaved plant named Sweet Basil. Dark Opal, an ornamental purple-leaved variety, has somewhat less flavor but can also be used. In either case, collect young leaves; dry them in an airy room; crumble them, and pack in tight containers.

BASS

One of the prime favorites of both fresh and salt-water anglers, the bass is excellent to eat. Wrap and freeze in the usual way. Store for 6–9 mo.

BAY LEAVES

The bay tree is a handsome, dense evergreen that grows in our warmest climates. The aromatic leaves are used to season soups, stews, sauces, fish, and pickles. Collect the young leaves when the tree is making growth, and dry them in any airy place. Package in tight containers or polyethylene bags.

BEACH PLUMS

Beach plums, which grow on the North Atlantic coast, are deciduous shrubs. In late summer they produce large quantities of small, usually purple fruits with a heavy bloom. These are used to make an outstanding jelly or jam, and are sometimes baked in pies.

JELLY

Makes 9 8-oz. glasses

2 qt. beach plums
2 1/2 qt. water
6 cups sugar
1/2 bottle liquid pectin

Wash, pit, and crush the fruit; place in a kettle with water; bring to a boil, and simmer for 30 min. Extract juice and measure out 3 1/2 cups. Combine with sugar and follow standard procedure for making jelly with liquid pectin.

Beach plum jelly can be made without added pectin—like plum jelly—but the fruit does not always contain sufficient natural pectin to give assured results.

JAM

Make it like plum jam.

FREEZING

Mix 3 cups washed, pitted plums with 2 cups sugar and 2 tsp. ascorbic acid. Pack in rigid containers and freeze. Store for 10–12 mo.

BEANS, FAVA

Fava beans are also called broad or horse beans. The bushy plants produce large, inedible pods containing big, flat seeds rather like lima beans and preserved and used in the same ways.

BEANS, LIMA

Varieties especially good for freezing and drying include Fordhook U.S. 242, Henderson's Bush, and Thorogreen.

FREEZING

Shell and discard starchy beans. Sort according to size. Follow standard freezing procedure. Blanch large beans 2 min.; medium beans, 1 1/2 min.; small beans, 1 min. Store for 10–12 mo. To serve, cook small beans 10–12 min.; medium beans, 12–14 min.; large beans, 16 min.

CANNING

Raw pack. Shell and wash beans. Pack loosely in jars. For small beans allow 1 in. head space in pints; 1 1/2 in. in quarts. For large beans allow 3/4 in. head space in pints; 1 1/4 in. in quarts. Add 1/2 tsp. salt to pints; 1 tsp. to quarts. Cover with boiling water to 1/2 in. of top. Seal. Process in pressure canner at 10 lb. pressure. Pints for 40 min.; quarts for 50 min.

Hot pack. Shell, wash, place in boiling water, and bring to a boil. Pack loosely in jars to 1 in. of top. Add 1/2 tsp. salt to pints; 1 tsp. to quarts. Cover with cooking water. Allow 1 in. head space. Seal. Process in pressure canner at 10 lb. pressure. Pints for 40 min.; quarts for 50 min.

DRYING

Use mature, tender beans. Shell and wash. Blanch in boiling water for 5 min. Drain and spread a 1/2-in. layer in trays. Put in 115° oven and increase heat gradually to 140°. Stir frequently. Beans should be dry in 6–10 hr. When cool, pack into tight containers and store in a cool, dark, dry place.

BEANS, SHELL

Any beans such as limas and string beans might be called shell beans if you allowed them to mature and dry and you then removed the fat seeds from the pods. But the real shell beans are those which are handled only in this way. They come in various shapes, sizes, and colors and are known by such names as Navy, kidney, pinto, marrowfat, and French Horticultural. They are usually baked, and according to New England tradition, they should be eaten every Saturday night—with brown bread, of course.

The old and easiest way to preserve shell beans is to let them dry on the vine (but take them in before they are hit by frost); then shell them, and store them in tight containers in a dark place at room temperature. To prevent a possible infestation of weevils, it's a good idea before packaging the beans to heat them for 30 min. in a 130° oven; then let them cool completely.

A newer and better drying method is to harvest the beans when they are at their peak. Blanch the seeds in boiling water for 5 min. Drain. Then spread in trays and dry in the oven for 6–10 hr. Start the oven at 115° and gradually increase to 140°. Stir the beans often.

FREEZING

Some shell beans, such as the French Horticultural, are reasonably good to eat green if harvested while they are succulent. To freeze, remove the seeds from the pods (which are inedible) and blanch for 1 min. Follow standard freezing procedure. Store for 10–12 mo. To serve, cook for 12–16 min.

BEANS, SNAP

FREEZING

Varieties especially good for freezing include Brittle Wax, Eastern Butterwax, Kinghorn Wax, Tendercrop, Tenderette, and Tenderpod. Select small, crisp beans. Wash and remove ends. Cut or break in 1-in. pieces, or French. Follow standard freezing procedure. Blanch for 2 min. Store for 10–12 mo. To serve, cook cut beans 12–15 min.; Frenched beans, 8–10 min.

CANNING

Varieties especially good for canning include Brittle Wax, Kinghorn Wax, Tenderette, and Tenderpod.

Raw pack. Wash beans, remove ends, cut or break in 1-in. pieces. Pack tightly into jars. Add 1/2 tsp. salt to pints; 1 tsp. to quarts. Cover with boiling water. Leave 1 1/2 in. head space. Seal. Process in pressure canner at 10 lb. pressure. Pints for 20 min.; quarts for 25 min.

Hot pack. Wash beans, trim off ends, cut or break in pieces. Place in boiling water and boil for 5 min. Pack loosely in jars. Add 1/2 tsp. salt to pints; 1 tsp. to quarts. Cover with cooking water to 1/2 in. of top. Seal. Process in pressure canner at 10 lb. pressure. Pints for 20 min.; quarts for 25 min.

BRINING

See Chapter 6.

DILLY BEANS

Makes 6 pints

4 lb. green beans
1 tbs. dill seed
18 peppercorns
6 tbs. salt
3 cups vinegar
3 cups water

Wash beans, remove ends and pack whole beans lengthwise into jars to 1/4 in. of top. Add 1/2 tsp. dill seed and 3 peppercorns to each jar. Combine salt, vinegar, and water and heat to boiling point. Pour over the beans, covering them. Seal. Process in boiling water bath for 20 min.

BEAR

A favorite recitation of Maine guides is titled "How to Cook a Bear." (Or sometimes it is moose meat or loon.) The details vary with the storyteller and occasion; but the gist of the account is this:

Let the meat age for a week. Then put it

in a marinade for several hours. Turn frequently. Transfer it to a moderate bake oven. Be sure to use nothing but birch for fuel. Roast for 4 hours. Then remove from oven, make a rich pie crust, and seal the meat in it. Return the meat to the oven and roast for another hour. Then put the meat on your best platter and garnish it with sprigs of watercress from the spring.

Carry the meat to the cabin door to call everybody in. Just as you yell, "Come and get it!" stub your toe on the doorsill. As you fall forward, pitch the meat as far as you can into the woods.

Bear meat, in other words, is a dubious delicacy. Still, if you shoot a bear, you can't be blamed for saving some of the meat so you can brag a bit to friends. Use the same preservation methods as for venison. The best cut is the loin, but other cuts can also be preserved.

BEECHNUTS

Beechnuts are small, brown, three-cornered nuts borne in pairs in prickly husks. They are produced by both American and European beeches, and are prized for their flavor. The nuts fall from the husks after the first frost. Gather them quickly—before the squirrels beat you; let them cure for several days, and then store them in sacks in a cool, fairly humid place.

BEEF

Beef must be thoroughly chilled immediately after slaughtering, and it usually is not allowed to hang more than a week before it is processed. Certain prime cuts are sometimes aged for several weeks, however.

FREEZING

Since beef is not improved by freezing (nor is it harmed), you should always start with high-quality cuts. There are no special tricks in handling. Simply wrap as tightly as possible in aluminum foil or other moisture-vaporproof material, label, and freeze. Ground meat should be stored no more than 4 mo. Large cuts can be held for 10–12 mo.

Beef kidneys, hearts, and tongues; beef and calf's liver; and calf's brains and sweetbreads are all easily frozen and stored for 4–5 mo. Wrapping in aluminum foil is generally most convenient, though rigid plastic containers are often used for brains and sweetbreads. Fresh tripe can be frozen in similar fashion.

CANNING

Follow standard procedure for canning meat and poultry in a pressure canner.

If beef is cut in strips or chunks, process pints for 75 min.; quarts for 90 min. The same time is used in both hot packing and raw packing.

Ground beef should be processed only by the hot-pack method. Use fresh meat, not a combination of fresh meat and cooked left-overs. Form into patties and bake in a 325° oven until almost no red color shows in the center. Drain off fat completely and dry patties with paper towels. Pack in clean, hot jars to 1 in. of top. Cover with boiling broth to 1 in. of top. Seal and process. Pints for 75 min.; quarts for 90 min.

CORNED BEEF

Use cheaper cuts for corning. Remove bone. Spread a layer of dairy salt in the bottom of a sterilized crock or similar container. Pack in a layer of meat. Sprinkle with salt; put in another layer of meat; and so on until all the meat is used up. Before packing, each piece of meat should be well rubbed with salt. In all you should use about 1 lb. salt for each 10 lb. meat.

Put the crock in a refrigerator and let it stand for 24 hr. Then cover with brine and weight the meat down. The brine is made of 6 1/2 oz. sugar, 1/2 oz. saltpeter, 1/4 oz. baking soda, and 6 1/2 cups water for each 10 lb. meat. Return the crock to the refrigerator and let the meat cure for 5 wk. Then pour off the brine and wash the meat in fresh water.

The corned beef can then be dried, wrapped, and frozen but it should be used within 4–6 wk. To can the beef, cut it in pieces, put in a kettle, cover with cold water, and bring to a boil. Then pack in jars to 1 in. of top and cover with the cooking water. Leave 1 in. head space. Seal and process in a pressure canner at 10 lb. pressure. Pints for 75 min.; quarts for 90 min. (If the water in which the corned beef is first boiled is too salty to tolerate, pour it off, cover the meat with fresh water, and bring to a boil again.)

DRIED BEEF

Cure the meat like corned beef. When curing is completed, pour off the brine, wash the meat in cold water, and hang it up to dry for 24 hr. Then smoke for 70 to 80 hr. at 130°. If meat is well dried out, it can be wrapped to keep off insects and hung in a dry, dark, well-ventilated place at normal temperatures.

JERKY

See *Venison.*

PICKLED TRIPE

Wash clean, white tripe and place it in a crock. Cover with a brine made in the proportions of 2 lb. salt to 1 gal. water. Let stand for 4 days in a cool place. Then pour off brine and wash tripe well in fresh water. Return to the cleaned crock and cover it with cider vinegar. Weight down and cover crock tightly. Store in a refrigerator for up to 4 mo.

LARD

Beef fat may be rendered and mixed in modest quantities with pork fat. See *Pork.*

BEETS

DRY STORAGE

Trim off all but 1/4 in. of the tops, pack in slightly moist sand, and store at 33° to 40°. The beets will keep for up to 3 mo.

FREEZING

Use beets not over 3 in. in diameter. Trim off tops and wash. Cook in boiling water until tender, roughly 30 to 45 min. Cool completely in cold water. Peel, slice, or dice. Small beets can be left whole. Pack in rigid containers to 1/2 in. of top, seal, and freeze. Store for 10–12 mo. To serve, heat in water till warm.

CANNING

Use beets not over 3 in. in diameter. Cut off tops and wash. Boil in water till skins slip easily, roughly 15 to 25 min. Peel, slice, or dice. Small beets can be left whole. Pack in jars to 1/2 in. of top. Add 1/2 tsp. salt to pints; 1 tsp. to quarts. Cover with cooking water. Seal. Process in pressure canner at 10 lb. pressure. Pints for 30 min.; quarts for 35 min.

PICKLED BEETS

Use beets not over 3 in. in diameter. Cut off tops and wash. Cook in boiling water until tender. Peel, slice, and pack in jars to 1/2 in. of top. Add 1/2 tsp. salt to pints; 1 tsp. to quarts. Meanwhile heat 2 cups cider vinegar and 2 cups water to boiling and pour this over the beets. Allow 1/2 in. head space. Seal. Process in boiling water for 30 min.

BEET GREENS

The leaves of all beets can be eaten as greens when young; but those of Green Top, Detroit Dark Red, and Long Season are particularly good. For how to freeze or can the tops, see *Greens.*

BIGNAY

The bignay is a shrubby tropical evergreen tree which bears clusters of blue-black fruit much like currants. The methods given here for preserving are adapted from those recommended to us by Olga M. Kent of Coconut Grove, Florida. Miss Kent's colleagues in the state Agricultural Extension Service, from which she is now retired, bill her as knowing more about the home processing of tropical fruits than anyone around.

FREEZING

Wash and stem ripe fruit, purée in a blender, and press through a sieve. Mix with sugar to taste. Pack in rigid plastic containers and freeze for use in iced drinks or as a sauce over ice cream. Store for 10–12 mo.

JUICE

Wash and stem ripe fruit, barely cover with water, and boil for 20 min. Extract juice in a jelly bag. Mix 3 cups with 1 cup sugar. Return to the range and bring to a boil; then pour into hot, sterilized Mason jars and seal. Process in boiling water for 5 min.

JELLY

Makes 9 8-oz. glasses

3 1/2 lb. bignays
7 cups sugar
1 bottle liquid pectin

Use semiripe red fruit with a few unripe yellow berries thrown in. Wash and stem fruit. Barely cover with water and boil for 20 min. Extract juice. Measure 4 cups into kettle and combine with sugar. Follow standard procedure for making jelly with liquid pectin.

BLACKBERRIES

FREEZING

Sugar pack. This is the best packing method if berries are to be cooked before serving. Use fully ripe but firm fruit. Mix 4 to 5 lb. with 1 lb. sugar.

Sirup pack. Use this method if you intend to eat berries without cooking. Pack fruit into rigid containers and cover with cold, 40 or 50 percent sugar sirup.

Blackberries can also be packed dry for cooking. Storage life in all cases is 10–12 mo.

CANNING

Raw pack. Use firm, ripe berries. Pack in jars to 1/2 in. of top. Shake berries down but don't crush them. Cover with boiling, 40 or 50 percent sugar sirup to 1/2 in. of top. Seal. Process in boiling water. Pints for 10 min.; quarts for 15 min.

Hot pack. Mix 1/2 cup sugar with each 4 cups washed fruit. Bring to a boil in a covered pan. Pack in jars to 1/2 in. of top. Seal. Process in boiling water. Pints for 10 min.; quarts for 15 min.

JELLY

Without added pectin
Makes 4 8-oz. glasses

2 1/2 qt. blackberries
3/4 cup water
3 cups sugar

Crush washed berries in a kettle, add water, cover, and bring to a boil. Simmer for 5 min. Extract juice and measure out 4 cups. Mix with sugar. Follow standard procedure for making jelly without added pectin.

JELLY

With added pectin
Makes 10 8-oz. glasses

3 qt. blackberries
1/4 cup lemon juice
1 bottle liquid pectin

Berries should be fully ripe. Wash and crush them, put in a jelly bag, and squeeze out juice. Measure 4 cups into a kettle. Add lemon juice and sugar. Follow standard procedure for making jelly with liquid pectin.

JAM

Without added pectin
Makes 4 8-oz. glasses

2 qt. blackberries
4 cups sugar

Crush ripe berries and measure 4 cups into a kettle. Mix in sugar. Follow standard procedure for making jam without added pectin.

JAM

With added pectin
Makes 9 8-oz. glasses

2 qt. blackberries
7 cups sugar
1/2 bottle liquid pectin

Crush ripe berries and measure out 4 cups. Add sugar. Follow standard procedure for making jelly with liquid pectin.

JUICE

Crush ripe berries in a kettle and heat to 175°. Extract juice and strain. Add 1 cup sugar to 9 cups juice if desired; but usually it is not necessary. Then follow standard juice-making procedure.

SIRUP

See Chapter 9.

BLOWFISH

The blowfish, or puffer, is a small saltwater fish with the rare ability of blowing itself up like a balloon when it fears attack. More important, it has such delicate, juicy meat that it is also called sea squab. The meat can be frozen in aluminum foil and stored for about 6 mos. *But don't try it.*

John A. Peters, of the federal government's Atlantic Fishery Products Technology Center, tells us: "The puffers (blowfish, swellfish, etc.) in general have viscera that are highly toxic. The degree of toxicity will vary with the species, but all should be treated with great respect. Only the skinned flesh back of the visceral cavity is edible, and precautions must be taken to be sure that all parts of the liver and other viscera are removed."

To reinforce this view, Dr. Bruce W. Halstead, director of the World Life Research Institute, notes in his book, *Dangerous Marine Animals,* that "eating puffer is a game of Russian roulette. It makes an excellent poisonous bait for stray cats, but a poor food for humans."

BLUEBERRIES

Traveler Mountain Pond in north central Maine used to be something special, and perhaps it still is. But we are not going to risk our pleasant dreams trying to find out.

We reached it in one of two ways. Sometimes we followed the ancient hazelnut-bordered logging road from the outskirts of Patten to Charlie McDonald's "sporting" camp on the East Branch of the Penobscot. Other times we hiked through the woods to Jerry Pond and Marble Pond, where the moose browsed in the poplars; and then on to Bowlin Pond and the wobbly fishing "throne" McDonald had built so that his sports could cast over the marsh grass to reach the big native trout in the pond's outlet; and finally we arrived at McDonald's camp.

The next morning, after a large breakfast of flapjacks, trout, toast, coffee, and chocolate doughnuts, we crossed the river and started the long climb up the mountain. Soon after lunch we reached the scrub growth; then the big patches of lowbush blueberries; then the pond itself.

The first time we made the trip, as we got perhaps a quarter mile above the woods, we spied a black bear gorging on the blueberries. Even in those days, we didn't come on bear very often because, although there were plenty of them, they usually heard us first. This time was an exception, but not for long. Within seconds of our sighting, he had heard or winded us and was off, lumbering over a ridge.

We went on to McDonald's lean-to, dropped our packs, and climbed to the pond. It was a tiny thing, only about a quarter mile in diameter. But the old raft we found at the end of the trail was an indication that fishing was good. Two of us climbed on; the others stayed ashore. No one was unhappy.

With the first cast of a Parmachenee Belle —or maybe it was a Royal Coachman—a trout was hooked and brought home. And so it went. The fish were everywhere and eager to bite, even in the early afternoon. We put two flies on the leader and promptly caught two trout at a cast. Three flies and the same thing. Within a short time we had caught more than we could use and put on barbless hooks just for the fun.

But the fishing was too good. Though the trout were small—not much over 9 inches long at most—they were just too numerous. The big fun remained just down the hill.

We took the trout back to camp and ventured forth to gather blueberries in our old red felt fishing hats from L. L. Bean. They were everywhere. People today are so used to the big fat blueberries that are grown on cultivated highbush plants that they don't realize what a good blueberry really is. We do not mean to disparage the cultivated highbush blueberries, because they are beautiful and delicious. But the fruit of a lowbush blueberry is not to be equaled. It's small, but it's plentiful and my, does it taste good! A bit sharp; not a bit soft and mealy; something to bite into and savor.

We forget how many hats we filled. When we finally called a halt, there were still plenty of berries left for Papa Bear and Mama Bear and Who-is-in-my-bed-little-Goldilocks Bear—and all their relatives. Anyway, we took them back to the lean-to and Charlie Gagnon, our guide, mixed part of them with flour and sugar and steamed them in an old rusty can he had found. And after we had eaten trout till we couldn't eat another bite, we ate blueberry steam pudding till we burst.

The next day, after fishing some more, we picked berries again. And that's when we got a little silly, because we lined one of our packs with newspaper and carried Lord knows how many blueberries back to home camp. These we made into blueberry jam.

It should have lasted a year. But we finished it off within two months.

FREEZING

Pick over, stem, and wash berries. Berries which are to be used in pies, puddings, and muffins are slightly better if packed in 40 percent sugar sirup. For eating, mix 4 to 5 lb. berries with 1 lb. sugar and package dry. If the skins are tough, you can tenderize them by placing the berries in boiling water for 1 min. before packing. Store for 10–12 mo.

CANNING

Pick over, stem, and wash berries. Mix 1/2 cup sugar with each quart fruit and bring to a boil, *shaking* often. Then pack in jars to 1/2 in. of top. Seal. Process in boiling water. Pints for 10 min.; quarts for 15 min.

SIRUP

Follow standard procedure for making juice and sirup.

ICE CREAM

See *Milk.*

JAM

Makes 11 8-oz. glasses

3 pt. blueberries
2 tbs. lemon juice
7 cups sugar
1 bottle liquid pectin

Wash, stem, and crush berries. Measure 4 1/2 cups into a kettle. Add lemon juice and sugar. Follow standard procedure for making jam with liquid pectin.

PRESERVES

Makes 5 half-pints

1/2 cup water
6 cups sugar
2 qt. blueberries

Boil water and mix in sugar until dissolved. Add washed and stemmed berries and boil, *shaking* frequently, until they are translucent when lifted from sirup, about 15 min. Then place in a flat pan and *shake* frequently so berries will take up sirup. Before they lose their heat, pack in Mason jars to 1/2 in. of top. Seal. Process in simmering water to cover for 15 min.

BLUEFISH

Bluefish come from the sea and are prized not only for the fight they put up but also for their good eating. Even the snapper blues—6 to 9 in. fish caught off the docks—are good fun. Whatever the size, wrap and freeze them in the usual way. Store for 6–9 mo.

Bluefish can also be salted or given a cold smoke.

BLUE RUNNERS

These 1- to 2-lb., lean fish are most often caught off the Florida coast. Freeze in aluminum foil. Store for 6–9 mo.

BONITO

The bonito is a large member of the mackerel family, very similar to the albacore. Freeze or can it like tuna. Or salt it according to standard procedures.

BOYSENBERRIES

These berries are a variety of blackberry. They are unusually large and of a purplish cast. Otherwise there is nothing unique about them. Put them up like blackberries.

BROCCOLI

Varieties especially good for freezing include Green Comet, Green Mountain, and Waltham 29.

Harvest large central heads and the small sprouts that follow before they flower. Soak in salt water to remove worms, if these are a problem. Cut lengthwise into pieces about 1 in. across. Follow standard freezing procedure. Blanch for 3 min. Store for 10–12 mo. To serve, boil for 4–6 min.

BRUSSELS SPROUTS

If you don't want to preserve Brussels sprouts for a long period, you can leave them in the garden for some time after the first fall frost, because they can withstand temperatures down to about 10° if not exposed to them too long. Long-term preservation is a simple matter, however.

DRY STORAGE

Dig up entire plants in the fall, stand them upright and close together in the storage room, and tamp moist soil around the roots.

FREEZING

Use firm, tight, dark green heads. Remove old outer leaves. Grade according to size. Follow standard freezing procedure. Blanch small heads for 3 min.; large heads, for 4 min. Store for 10–12 mo. To serve, boil for 5–7 min.

BUCKWHEAT

Buckwheat is not a member of the grass family and is therefore not related to wheat or other cereal grains. It is, however, used to make flour for buckwheat pancakes, and if you have never eaten real buckwheat pancakes—not the kind that comes out of a supermarket box—you had better find yourself a source of buckwheat right now.

Buckwheat is a 3-ft. plant with heart-shaped leaves and fragrant flowers which the bees tap to make buckwheat honey. Each of the flowers produces a single black or gray, three-sided seed.

Harvest the seeds just before they are fully dry and let them continue drying in a warm, airy place. To turn them into flour, grind them through a grist mill.

BUFFALO BERRIES

The Indians of the Northern Plains States used to dry buffalo berries, and you can, too. Just spread them out in a sunny spot or put them in the oven. But the 1/4-in., very sour berries, which are borne on a silvery-leafed shrub that thrives in the Dakota Badlands, deserve better things.

JELLY

Makes 4 8-oz. glasses

1 1/2 lb. buffalo berries
2 1/2 cups water
2 1/4 cups sugar

To pick the berries without being wounded by the spines on the plant, wait until after a hard frost. Then put a blanket under the shrub and hit the branches with a stick.

Wash the berries thoroughly, pick out leaves, and remove as many stems as possible. Cover with water and boil slowly for 10 min. Crush fruit and extract juice, which has a disagreeable odor and therefore should not be put in a refrigerator. Let drip overnight; then bring to a boil again and re-strain through your jelly bag. Measure out 3 cups juice, combine with sugar and boil to just before the jelly stage. Ladle into hot, sterilized glasses and seal. The jelly becomes firmer in storage. Keep in a dark closet, because it fades in the light.

BUFFALOFISH

This is a big, lean-meated fish often caught in the rivers of the Mississippi Valley. Freeze whole fish, steaks or fillets in the normal way. Store for 6–9 mo.

Buffalofish are also smoked and then frozen.

BURBOT

Burbots are codfish growing in our northernmost fresh waters. The fish are slender and green, the meat lean and delicate. Wrap in aluminum foil and freeze for 7–10 mo.

BUTTERFISH

The butterfish swarms in the Atlantic, is about the same size and shape as a silvery butter plate, and makes very good eating when pan-fried. Wrap in aluminum foil and freeze. Store for 6–9 mo.

The fish can also be given a hot smoke.

BUTTERNUTS

The butternut is a species of walnut and has a general resemblance to the black walnut except that the wood is white and the rough, thick-shelled nuts are smaller. Harvest and preserve the nuts like black walnuts. If they prove very difficult to crack, try covering them with boiling water for 10 min.

CABBAGE

Cabbage can be frozen and canned, but it is then suitable only for cooking and not very good at that. So don't waste your time with these preservation methods.

The oldest and simplest way to keep cabbage is to put it in a dark, humid, dry-storage area at 33° to 40°. Spread the heads out a little so that air can circulate around them, but don't expose them to direct air movement.

SAUERKRAUT

Makes 7–8 quarts

25 lb. cabbage
5/8 cup kosher or dairy salt

Use firm, ripe cabbage. Let them stand for a day at room temperature so they will wilt and be easier to pack. Trim off outer leaves. Wash. Cut in quarters and remove cores. Cut into shreds no thicker than a dime. Mix the shredded cabbage and salt together thoroughly in a large container and let stand 3 min. or a little longer.

Pack part of the cabbage into a large, clean crock and push it down until the juice shows on the surface, Continue in this way until the crock is full to within about 3 in. of the top. Pour a pound or two of water into a large plastic bag, seal the top with a string, and drop it into the crock on top of the cabbage. If necessary, spread out the water cushion so that it presses against the sides of the crock. This not only weights down the cabbage under the brine but also keeps the air from reaching it. Thus it prevents growth of molds.

Keep the crock in a room at a temperature close to 70°. Remove scum daily. Gas bubbles that form indicate that the sauerkraut is fermenting. Fermentation should be completed within 4 to 6 wk. When this happens, put the kraut in its own juice in a kettle and heat to simmering (about 190°). Then pack the cabbage into clean, hot jars to 1/2 in. of the top. Cover with the juice. If there is not enough juice, make a brine of 2 tbs. salt and 1 qt. hot water. Leave 1/2 in. head space. Seal.

Bring water to a full boil in your canning kettle. Immerse the hot jars and start timing at once. Process pints for 15 min.; quarts for 20 min.

CALAMONDINS

Calamondins are small citrus fruits resembling tangerines but with a very acid pulp.

MARMALADE

Makes 3 8-oz. glasses

4 calamondins
3 cups sugar

Wash, halve, remove seeds, and slice fruits thin. Measure into a kettle and add 3 cups water for each cup fruit. Bring to a boil and simmer for 15 min. Measure out 3 cups pulpy juice and combine with sugar. Cook to jam stage. Remove from heat and let cool to 190°. Ladle into hot, sterilized glasses and seal.

PRESERVES

Follow directions for preserving kumquats.

CANDIED CALAMONDINS

Follow directions for candying grapefruit. Calamondins are candied whole, however. Just wash them and make a small cut across each fruit. Only one boiling is needed to remove bitterness.

CARAMBOLA

This is a small tropical evergreen tree that bears large, waxy, yellow fruits which are star-shaped in cross-section. To freeze, peel ripe fruits and slice crosswise. Pack in rigid containers and cover with 40 percent sugar sirup. Freeze. Store for 10–12 mo.

CARAWAY

Caraway is a biennial plant which doesn't set seeds until the second summer after planting. When these are dry, store in bottles for use in breads, cookies, cakes, potato salad, and baked fruit.

CARDOON

This large perennial vegetable is a close relative of the artichoke and is grown in the same part of the country for its coarse, celerylike leaf stalks. Harvest the mature stalks after they have been blanched in heavy paper or black plastic. Remove the leaves, wash, and scrape the stalks with a knife and cut them into 2- to 3-in. lengths. Follow standard freezing procedure. Blanch in boiling water for 4 min. Store for 10–12 mo. To serve, boil for 5–8 min.

CAROB

The carob, or St. John's-bread, is a large, evergreen tree grown in warm climates—mainly California. It bears foot-long, seedy pods which are rich in food value, though they are more commonly consumed by cattle than human beings.

In the autumn, the pods turn dark brown and fall to the ground. Those that remain in the tree can be knocked down. Allow them to dry for several days under the tree. They can then be stored in burlap bags in a dry, well-ventilated place for many months.

To make into a chocolate-flavored meal, crumble the dry pods by hand or put them through a food grinder or grist mill. Store the meal in tight containers in a dark place.

CARP

Package and freeze this fish in the usual way. Store for 6–9 mo. Carp also makes a good smoked product.

CARROTS

DRY STORAGE

Dig up carrots, remove tops, and place the roots without washing on a 1-in. layer of sand in the bottom of a box. Cover with sand; make another layer of carrots; and so on until the carrots are used up. Keep the box in dry storage at just above freezing. They will keep for 4–5 mo.

FREEZING

Varieties especially good for freezing include Danvers Half-Long and Scarlet Nantes. Use small, young carrots. Remove green flesh at the top. Wash, scrape, and cut the roots into 1/4-in. slices. Follow standard freezing procedure. Blanch for 2 min. Store for 10–12 mo. To serve, cook for 5–10 min.

CANNING

Hot pack. Wash, scrape, and cut carrots into 1/4-in. slices. Boil for 3 min. Pack without crushing in jars to 1/2 in. of top. Add 1/2 tsp. salt to pints; 1 tsp. to quarts. Cover with cooking liquid to 1/2 in. of top. Seal. Process in pressure canner at 10 lb. pressure. Pints for 25 min.; quarts for 30 min.

Raw pack. Wash, scrape, and cut into 1/4-in. slices. Pack tightly into jars to 1 in. of top. Add 1/2 tsp. salt to pints; 1 tsp. to quarts. Cover with boiling water to 1/2 in. of top. Seal. Process in pressure canner at 10 lb. pressure. Pints for 25 min.; quarts for 30 min.

BRINING

See Chapter 6.

CATFISH

Succulent channel cats no longer abound in southern and midwestern rivers, but the huge catfish farms of Mississippi, Arkansas and other states with warm climates have taken up the slack. Today, as a result, the United States may have a larger catfish population than ever before. And for a lot of people, that is good news.

Wrap and freeze catfish in the usual way. Store for 4–6 mo. The fish can also be salted.

CATNIP

Catnip, or catmint, is a perennial plant with mildly mint-flavored leaves. Collect, dry, and crumble these before the flowers open, and use them to make a flavorful tea, which is said to be good for colds and headaches. Let the leaves steep for 10 min. or more; don't boil them.

The dried leaves will also delight your cat.

CAULIFLOWER

Cauliflower is an important ingredient of mustard pickles (see *Cucumbers*) and chowchow (not described). But it is most commonly preserved by freezing.

Use solid heads that have not begun to crumble. Break or cut the heads into pieces no more than 1 in. thick. Soak in salt water for 30 min. if necessary to remove insects. Follow standard freezing procedure. Blanch for 3 min. Store for 10–12 mo. To serve, cook for 5–8 min.

CELERIAC

Celeriac is a variety of celery which is grown for its knobby, celery-flavored root.

This is sometimes boiled and eaten as a vegetable or cut up and used in stews and soups. To store, cut off the tops and place the roots, without washing, in a box of sand in your dry storage area at 33° to 40°.

CELERY

Celery is blanched white only for the sake of appearance. The process does not affect the flavor or texture, but it does reduce keeping quality.

The vegetable is an ingredient of several pickles described under *Cucumbers* and *Corn.*

DRY STORAGE

Dig up the entire plant, stand it upright in the storage area at 33° to 35°, and pack moist sand or soil around the roots. The stalks will remain in good condition for 2–3 mo.

FREEZING

Use crisp, solid stalks free of coarse strings. Trim, wash, and cut into 1-in. lengths. Follow standard freezing procedure. Blanch for 2 1/2 min. Store for 10–12 mo. To serve, cook for 12–15 min.

CANNING

Wash and cut crisp stalks into 1-in. lengths. Boil for 2 min. Pack into jars to 1/2 in. of top. Add 1/2 tsp. salt to pints; 1 tsp. to quarts. Cover with cooking liquid. Seal. Process in pressure canner at 10 lb. pressure. Pints for 20 min.; quarts for 25 min.

BRINING

See Chapter 6.

DRIED LEAVES

Use for seasoning. Cut off leaves and dry them in a cool place out of the sun or in a warm oven. When crisp, crumble into bits and store in tight containers.

CELERY SEED

Harvest seed heads just before the seeds are completely dry, and let them continue drying on cheesecloth in a cool place out of the sun. Then shake out the seeds and pack them into bottles.

CELTUCE

This is a lettuce relative with leaves which are used in salads and thick stalks which are boiled as a vegetable. They taste like a cross between lettuce and celery. To freeze the mature stalks, remove the leaves, wash, and cut into 1-in. lengths. Follow standard freezing procedure. Blanch for 2 1/2 min. Store for 10–12 mo. To serve, cook for 12–15 min.

CHAYOTE

Chayote is a large, perennial vine growing mainly in warm climates. It bears small, pear-shaped, green, squashlike fruits with a single big seed. Put these up by freezing or canning, like summer squash—but without the seed, of course.

CHERRIES, SOUR

If you are not going to process sour cherries right away, pick them with their stems on so they will not lose juice.

FREEZING

Wash, stem, pit, and mix 4 to 5 lb. cherries with 1 lb. sugar. Store for 10–12 mo.

CANNING

Raw pack. Wash, stem, and pit cherries. Shake them down into jars to 1/2 in. of top. Cover with 30 or 40 percent sugar sirup. Allow 1/2 in. head space. Seal. Process in boiling water. Pints for 20 min.; quarts for 25 min.

Hot pack. Wash, stem, and pit cherries. Mix 4 cups fruit with 1/2 cup sugar and a very little water. Bring to a boil in a covered saucepan. Pour into jars to 1/2 in. of top. Seal. Process in boiling water. Pints for 10 min.; quarts for 15 min.

DRYING

Wash, stem, and pit fruit and drain it for 1 hr. Place in trays in a single layer and dry in the oven till leathery but not sticky. Start at 120° and increase gradually to 150°.

ICE CREAM

See *Milk.*

JELLY

Makes 10 8-oz. glasses

3 lb. cherries
1/2 cup water
7 cups sugar
1 bottle liquid pectin

Wash, stem, pit, and crush fully ripe cherries. Place in a kettle with water, bring to a boil and simmer, covered, for 10 min. Extract juice. Measure out 3 1/2 cups juice, mix with sugar, and bring to a boil. Follow standard procedure for making jelly with liquid pectin.

JAM

Makes 10 8-oz. glasses

3 lb. cherries
7 cups sugar
1 bottle liquid pectin

Wash, stem, pit, and chop cherries fine. Measure 4 cups into a kettle and add sugar. Follow standard procedure for making jam with liquid pectin.

PRESERVES

Makes 4 8-oz. glasses

3 lb. cherries
4 cups sugar

Wash, stem, and pit cherries and weigh out 2 lb. Drain off juice, combine it with sugar (if there is not enough juice to dissolve sugar, add a little water), and cook till sugar dissolves. Let cool. Add cherries and boil until they are glossy, about

15 min. Cover and let stand overnight in a cool place. Bring to a boil again in the morning and boil hard for 1 min. Immediately pour into hot, sterilized glasses and seal.

BRANDIED CHERRIES

English Morello is a particularly good variety for brandying, but you can use any other kind. Wash cherries; do not pit. Mix 1 lb. fruit with 1 lb. sugar. Put in a crock or glass jars and cover with brandy. Cover tightly and store in a cool place. Don't use for at least a week—preferably for 2 mo.

WINE

Pick over cherries, discarding unripe, damaged, and blighted fruit. Remove stems. Wash thoroughly to remove fungicide. Place in primary fermenter, crush as much as possible, and cover with water. Add 1 lb. sugar and 1/4 package dry wine yeast for each gallon of the mixture. Proceed as in Chapter 16. Add extra sugar after first fermentation if the juice is very sour.

CHERRIES, SWEET

FREEZING

Wash, stem, pit, and pack cherries into rigid containers. Cover with 40 percent sugar sirup to which ascorbic acid has been added. (Use 4 tsp. ascorbic acid per 4 cups of sirup.) Freeze. Store for 10–12 mo.

CANNING

Raw pack. Wash cherries and pit them or not, as you wish. Shake them down into jars to 1/2 in. of top. Cover with 30 or 40 percent sugar sirup to 1/2 in. of top. Seal. Process in boiling water. Pints for 20 min.; quarts for 25 min.

Hot pack. Wash and pit, or not. Add 1/2 cup sugar to 4 cups fruit. Add a little water to unpitted cherries to prevent sticking. Bring to a boil. Pack into jars to 1/2 in. of top. Seal. Process in boiling water. Pints for 10 min.; quarts for 15 min.

ICE CREAM

See *Milk.*

JAM

Makes 10 8-oz. glasses

3 lb. cherries
1/4 cup lemon juice
7 cups sugar
1 bottle liquid pectin

Wash, stem, and pit fruit, and chop fine. Measure 3 3/4 cups into a kettle. Add lemon juice and sugar. Follow standard procedure for making jelly with liquid pectin.

PRESERVES

Makes 7 8-oz. glasses

2 1/2 lb. cherries
5 cups sugar
1/3 cup lemon juice
1/2 bottle liquid pectin

Wash, stem, and pit ripe cherries. Measure 5 cups into a kettle, add sugar, and let stand for 30 min. Stir occasionally. Then bring to a boil, stirring. Remove from burner and let stand for 4 hr. Add lemon juice. Bring to a rolling boil and boil hard for 2 min., stirring constantly but not vigorously. Remove from heat and stir in pectin immediately. Stir and skim for about 10 min. Then pour into hot, sterilized glasses and seal.

MARMALADE

Makes 4 8-oz. glasses

1 orange
2 lb. cherries
1/4 cup lemon juice
3 1/2 cups sugar

Don't peel orange; chop the whole thing very fine. Barely cover with water, boil till soft, and then cool. Pit cherries and add with lemon juice and sugar to the orange. Bring to a boil, stirring to dissolve sugar. Boil rapidly until marmalade passes jelly test—about 30 min. Ladle into hot, sterilized glasses and seal.

JUICE

Wash, stem, pit, and chop cherries fine. Heat to 160°. Place in jelly bag, squeeze out juice, and strain it through washed cheesecloth. Add 1 cup sugar to 9 cups juice if desired. Then process juice according to directions in Chapter 9.

For a tastier, tart product, mix about 1 cup sour cherries with 4 or 5 cups sweet cherries.

CHESTNUTS

When the burs on chestnuts open in the fall, the nuts fall to the ground. They should be picked up every day because otherwise they will deteriorate badly. To cure the nuts, spread them out in a shady, airy place until they feel a little soft. This may take 3 to 7 days. Then store the nuts in perforated plastic bags in the fresh-food section of your refrigerator.

FREEZING

Make 1/2-in. cuts in the shells on the flat side. Put the nuts in boiling water for 10 min. Then peel off the shells and inner skins with a knife. Let the meats cool thoroughly. Pack in rigid containers or polyethylene bags and freeze. Store for 9–10 mo.

Chestnuts may also be puréed after boiling for about 15 min. or until soft. Cool by placing pan in cold water. Then pack in rigid containers and freeze.

DRYING

If you leave chestnuts in their shells in a dry room, they will eventually become dehydrated. To refresh them, just break off the shells and soak the kernels in water.

A faster drying method is to slash the shells and boil the nuts for 10 min. Then peel off shells and skins. Place in trays in a 150° oven and dry until shriveled and hard.

CHESTNUTS IN SIRUP

Makes 5 half pints

2 lb. chestnuts
3/4 lb. sugar
2 cups water
1/2 tsp. vanilla flavoring

Shell and skin chestnuts as described above. Simmer the kernels in water till nearly tender. Meanwhile boil sugar and water to make a sirup. To this add drained chestnuts and vanilla. Cook for 30 min. Ladle nuts into hot, sterilized jars. Continue boiling sirup if it is not yet thick. Then pour it over nuts to top of jars and seal.

CHICKENS

We got into chicken raising at the end of World War II by a fluke. Stan wrote a story in General Electric's employee magazine about the chicken brooders being turned out by a tiny, often overlooked division of the company; and in gratitude for this rare publicity, the division manager gave him one of the brooders.

Now, when you are young and poor and have a naturally thrifty streak and like to dabble in farming, what do you do with such a gift? Use it, of course—even if you don't know the first thing about raising chickens.

The coop we built was 12 ft. long, 4 ft. wide, and 1 ft. high. The wire mesh floor was raised off the ground about 30 in. For some forgotten reason—probably because we didn't know how hardy chicks are—we put the coop in the garage. Then we sent off to an up-country farm for 100 day-old White Leghorn cockerels. We chose Leghorns because they are too small to be grown commercially for meat; consequently, we were able to buy the cockerels for 2 cents apiece.

We'll spare you the details. Enough to say that the chicks arrived and thrived—all one hundred of them. And soon it became apparent that, even though cold weather had arrived, the birds had to be moved outdoors.

That was a day to remember. The site chosen for the coop was on top of a hill which could be reached from the garage only by making a circuit of almost our entire acre. Fortunately, Elizabeth's younger sister Mary was visiting us. So we put one end of the chicken-filled coop on a wheelbarrow, which Mary and Elizabeth pulled; Stan picked up the other end, and off we went.

It was slow going. Then as we reached the gate at the end of the driveway, something happened to the wheelbarrow. The coop tilted, and suddenly White Leghorn cockerels were pouring out of the open top and flapping off over the countryside.

Somehow we managed to retrieve them all. We completed the journey, and from that day on we considered ourselves experienced chicken growers. This is not to say we didn't have occasional problems. But outside of a fox that managed to open the coop and carry off a couple of birds, the only serious problem arose on the day Stan decided to kill chickens so that they could be dry-plucked by thrusting a knife through the mouth and into the brain. To this day we don't know whether he succeeded. As is the way with chickens, the first one had such violent reflex actions that Stan was sure he had failed. And being completely undone, he used an ax from then on.

Our first year with the cheap Leghorns turned out so well that we switched to much meatier New Hampshires. And eventually we went into egg as well as meat production. We continued the project for 6 years.

We still think back happily about the money we saved and, more important, the splendid food we packed into our freezer. Our broilers, roasters, and stewers were heavier and more tender than anything on the market. Of course, they wouldn't compare with market chickens today, because feeds have been improved so greatly. But this doesn't mean that if we started to raise chickens again today, using modern feeds, we couldn't do as well as the huge farms in Georgia. So could you.

FREEZING

Pluck, draw, and wash the chicken. Cut off head and feet. Package chicken whole, split in half, quartered, or completely dismembered. Wrap in aluminum foil or put in a polyethylene bag. If giblets are packed with the chicken, wrap them in wax paper, plastic, or foil. Freeze. Store for 10–12 mo.; however, if giblets are with the chicken, storage time should be reduced to 3 mo.

If you freeze livers or mixed giblets separately, pack them in rigid plastic containers. If a container isn't filled, fresh livers can be added as you accumulate them. Store livers for 2–3 mo.

CANNING

Follow standard procedure for canning poultry. Cut chicken into pieces; trim off lumps of fat. Remove bones if you wish, but not the skin. Pack loosely into jars with thighs and drumsticks next to the glass; breasts in the center; small pieces where they fit. When unboned chicken is processed by the hot-pack method, process pints for 65 min.; quarts for 75 min. Boned chicken processed by the hot-pack method is processed in pints for 75 min; in quarts for 90 min. Use the same times when processing by the raw-pack method.

If preserving giblets, separate livers from gizzards and hearts so that the flavors do not blend. Place in a saucepan, cover with hot water or hot chicken broth, and cook till medium done. Pack at once into hot, clean pint jars to 1 in. of top. Cover with boiling broth or water to 1 in. of top. Seal. Process in pressure canner at 10 lb. for 75 min.

SMOKING

Follow directions for smoking turkey.

CHICORY, MAGDEBURGH

Magdeburgh chicory is related to witloof chicory but is grown only for its long, tapering root, which is used as a coffee substitute.

Harvest the roots in the fall; wash thoroughly; scrape; cut into lengthwise slices; and blanch them in boiling water for 5 min. Then roast them in a 300° oven until dry and brittle. Put through a grinder and store in tight containers in a cool, dark place.

CHICORY, WITLOOF

Also called Belgian endive and French endive, witloof chicory is a slender, crisp, leafy plant which is used in salads like lettuce or braised and eaten as a vegetable. The plants are grown from seed sown in the spring. In the late fall, before the ground freezes, dig them up and cut off the tops within an inch of the roots. Sort the roots according to length and put them in a box filled with damp sand or soil. Store in a cool, damp place.

You can take the roots out of storage at any time for the next 5 or 6 months to force them into growth. Combine those of similar length. If not of similar length, trim off the bottoms. The best lengths for forcing are 8 to 12 in.

Put an inch of sand in the bottom of another box and stand the roots, top side up, on the sand. The roots can almost touch. Fill sand in around them and cover the tops with 6 in. of sand. Water thoroughly, cover the box with damp burlap, and store it anywhere at a temperature of 50° to 60°. Check the box every few days to make sure the sand is still damp. If not, sprinkle it lightly.

After about 3 weeks, the leaves will start to come up through the sand. Check the box daily, and as soon as the leaves appear, remove the sand and cut the leaves off just above the roots for serving. Then replace the sand, water it again, and force the roots a second time. After the second harvest, throw the roots away.

People who really dote on chicory keep a supply coming throughout the fall, winter, and spring. The best way to do this is to force about 6 to 12 roots at a time.

CHINESE CABBAGE

Chinese cabbage is not a true cabbage and has a considerably milder flavor. It resembles cos lettuce or very large, fat heads of witloof chicory. It can be frozen but is then useful only as a cooked vegetable. Trim off outer leaves and cut the heads into thin wedges. Blanch for 1 1/2 min. Follow standard freezing procedure. Store for 10–12 mo. To serve, boil for 8–10 min.

CHIVES

Chives are perennial plants which can be grown from seeds, but it's easier to start with established clumps. Grow them outdoors from spring until just before the first fall frost. Then dig them up, plant in pots, and let them grow on in a cool, sunny window. If you keep them watered and fertilized occasionally, they will supply you throughout the winter and can be moved outdoors again in early spring.

FREEZING

Wash chives and carefully shake off water or pat it off with paper towels. Cut into short pieces with scissors; spread out on a cookie sheet or aluminum foil. Put in the freezer until hard. Then package in polyethylene bags and return to freezer. Store for 6–8 mo. When thawed, the chives will be limp, but will retain their flavor.

CHOKECHERRIES

JELLY

Makes 10 8-oz. glasses

3 1/2 lb. chokecherries
3 cups water
6 1/2 cups sugar
1 bottle liquid pectin

Wash and stem fruit. Place in a kettle with water, bring to a boil, and simmer for 15 min. Extract juice. Measure out 3 cups and mix with sugar. Follow standard procedure for making jelly with liquid pectin.

CHUB

We used to call chub a trash fish and took pains to throw them into the woods, far from the lakes and turbid streams in which we caught them. So imagine our surprise to learn that chub from the Great Lakes is smoked and sold by the millions of pounds by commercial fisheries. You can handle those you catch in the same way. Follow directions in Chapter 4.

CITRON

There are two citrons. One is a variety of watermelon which is often used to make watermelon rind pickles. The other is a citrus fruit resembling a large lemon. This has a very thick, fragrant peel and very little pulp. The peel is candied like grapefruit peel.

CLAMS

FREEZING

Scrub hard in water to remove sand and bits of shell. Open with a knife over a pan in which the juices are caught. Snip off necks of little neck clams. Wash meats in a solution of 2 qt. cold water and 1 tbs. salt. Pack whole or ground into rigid plastic or glass containers. Cover with strained clam juice or with brine to 1/2 in. of top. Put a wad of crushed plastic on top to keep clams submerged. Place in freezer. Store for 6–9 mo.

CANNING

Clean, open, and wash meats in brine. Bring 1 gal. water and 1/2 tsp. citric acid powder from the drugstore to a boil, and drop in clams for 1 min. Drain. Pack whole or ground in pint jars—about 1 1/2 cups meat to a jar. Cover with hot, strained clam juice to 1/4 in. of top. Seal. Process in pressure canner at 10 lb. pressure for 70 min.

SMOKING

See Chapter 4.

JUICE

Simply reserve juice you save from clams used for soup, etc.; strain it, pour into rigid containers, and freeze. Store for 4–6 mo.

PICKLED CLAMS

Follow directions under *Oysters.*

COCONUT

Coconuts are harvested when fully grown but still green if you want the milk. For the firm, white meat, let the nuts ripen on the tree until the husks start to dry and turn brown. The nuts should then be cut down or allowed to fall naturally; in the latter case, however, they may be somewhat overripe. Once harvested, the nuts should be opened rather promptly because they tend to become rancid in storage.

Cut open the husk and then the hard nut shell. Remove the meat and peel off the brown skin. The meat can be left in pieces or grated on a hand grater. One nut yields about 2 tightly packed cups of grated coconut. Package in rigid plastic containers or polyethylene bags, and freeze. If you want a sweeter product, mix 1 part sugar with 10 parts coconut meat. Store for 6–8 mo.

Another way to preserve the nut meat is to split the nut in half and dry the meat in the sun or an oven until it separates from the shell. Then remove the meat from the shell and continue drying until it is hard and crisp, breaks with a snap, and has a uniform pearly white color. This is copra. It can now be grated, packed into sealed containers, and stored at room temperature for several months. Or if you have a press of some sort, you can squeeze out the oil for cooking and making soap.

COD

The flesh of the cod is lean and flaky. It's also a bit bland, but New Englanders (among others) have doted on it for generations and are particularly fond of young cod, which is called scrod. The easiest way to put up cod of any age or size is by wrapping in aluminum foil and freezing. It can be held for 7–10 mo.

Cod is also salted (see Chapter 5) or cold-smoked (Chapter 4).

COLLARDS

Collards are closely related to cabbages and used as a substitute for them by southerners. The plants grow to 3 ft. tall and form a cluster of crumpled leaves or a loose head. For how to preserve, see *Greens.*

COOT

The coot is an aquatic game bird with purplish meat. Draw the bird as soon as possible after shooting and place it where it can cool rapidly. Let it age in the refrigerator or a cool place for 24 to 48 hr. Don't pluck the bird; skin it. The whole bird can then be wrapped in aluminum foil or placed in a polyethylene bag and frozen. You can save freezer space, however, by cutting the meat from the breast and taking off the legs, and packaging these in foil, a bag, or a rigid plastic container. Store for 6–9 mo.

CORIANDER

Coriander is an annual herb grown for its orange-flavored seeds. Collect these in bottles when dry and use them in baked goods and salad dressings.

CORN, FIELD

MEAL

Leave corn in the field until ears are dry. Then pick, shuck, and shell the ears and put the kernels through a grist mill. Store the meal in glass or plastic containers or bags in a dark place.

HOMINY

Combine 2 qt. shelled, dry corn with 8 qt. water and 2 oz. lye. Boil hard for 30 min. and let stand off the heat 20 min. more. Drain and rinse the corn several times under hot water and then under cold water. Remove the dark tips of the kernels by working your hands through the corn for 5 min. or more. Place the corn in water to cover so that the tips will float to the surface and can be skimmed off. Then pour water over the corn until the surface is covered 1 in. deep, and boil for 5 min. Drain. Repeat this process four more times. Then boil the corn for 45 min., or until soft. Drain.

Pack the corn in clean, hot jars to 1/2 in. of top. Add 1/2 tsp. salt to pints; 1 tsp. to quarts. Cover with the boiling cooking water to 1/2 in. of top. Seal. Process in pressure canner at 10 lb. pressure. Pints for 60 min.; quarts for 70 min.

HOMINY GRITS

Proceed as above until the corn is cooked soft. Then drain and dry as well as possible on paper towels. Spread in 1/2-in. layers in trays and dry in the oven until hard and brittle. Start at 130° and gradually increase to 165°. When the hominy is cool, grind it in a grist mill and store in glass or plastic containers or bags.

CORN, SWEET

FREEZING

All yellow corns are good for freezing but Golden Cross Bantam is outstanding. Miniature is especially recommended for freezing on the cob.

Cut corn. Process as soon as possible after harvesting. Kernels should be full and yellow, and should spurt a thin milk when pressed with your fingernail. Remove all bits of husk and silk; cut out spots chewed by borers (also the borers). Follow standard freezing procedure. Blanch for 5 min. Chill thoroughly in cold water and then cut off the top two-thirds of the kernels with a sharp knife. Seal in rigid containers. Store for 10–12 mo. To serve, cook for 2–4 min.

Cream style. This is the best way to put up corn that has progressed beyond the milk stage. Handle as above, but cut off only the top half of the kernels. Then scrape what remains with the back of a knife to remove the thick juice.

On the cob. The results of this method are a little disappointing, since the corn doesn't taste quite like fresh corn on the cob. You also waste freezer space. But if you insist—pick young ears in the milk stage, remove all silks and bits of husk. Blanch small ears (less than 1 1/4-in. diameter) for 7 min.; medium ears (1 1/4- to 1 1/2-in. diameter) for 9 min.; large ears (over 1 1/2-in. diameter) for 11 min. Cool until cob is cold. Pack small ears in rigid containers or wrap large ears individually or in pairs in aluminum foil. Store for 10–12 mo. To serve, cook for 4–6 min.

CANNING

Whole kernel, raw pack. Use ears in milk stage. Husk and remove silks. Cut off kernels at about two-thirds of their depth. Pack loosely in jars to 1 in. of top. Add 1/2 tsp. salt to pints; 1 tsp. to quarts. Cover with boiling water to 1/2 in. of top. Seal. Process in pressure canner at 10 lb. pressure. Pints for 55 min.; quarts for 85 min.

Whole kernel, hot pack. Husk, desilk, and wash ears in milk stage. Cut off kernels at about two-thirds of their depth. Place 4 cups corn in 2 cups boiling water and bring to a boil. Strain off water and pack kernels in jars to 1 in. of top. Add 1/2 tsp. salt to pints; 1 tsp. to quarts. Cover with the cooking water to 1 in. of top. Seal. Process in pressure canner at 10 lb. pressure. Pints for 55 min.; quarts for 85 min.

Cream style, raw pack. Use mature corn. Remove husks and silks. Wash. Cut off kernels at about half their depth and then scrape cobs with the back of a knife. Pack loosely in pint jars to 1 1/2 in. of top. Add 1/2 tsp. salt. Cover with boiling water to 1/2 in. of top. Seal. Process in pressure canner at 10 lb. pressure for 95 min.

Cream style, hot pack. Remove husks and silks. Wash. Cut off kernels at about half their depth and scrape the cobs. Place 4 cups corn in 2 cups boiling water and heat to boiling. Pour into pint jars to 1 in. of top. Add 1/2 tsp. salt. Seal. Process in pressure canner at 10 lb. pressure for 85 min.

DRYING

Harvest during milk stage. Husk and remove silks. Blanch in boiling water until juice does not escape when kernels are cut. This takes about 10 min. for mature ears; up to 15 min. for very young ears. Cut off kernels at about two-thirds of their depth. Spread in trays in 1/2-in. layers and heat in oven for 6 to 10 hr., until hard and brittle. Start at 130° and increase gradually to 165°. Store in tight containers or plastic bags in a cool, dry, dark place.

BRINING

This method, described in Chapter 6, is used for corn on the cob.

CORN RELISH

Makes 7 pints

8 cups corn kernels (about 18 ears)
2 cups chopped sweet red peppers (4 to 5 medium peppers)
2 cups chopped green peppers (4 to 5 medium)
4 cups chopped celery (1 large bunch)
1 cup chopped onions
1 1/2 cups sugar
1 qt. vinegar
2 tbs. salt
2 tsp. celery seed
2 tbs. dry mustard
1 tsp. turmeric

Boil ears of corn in water for 5 min., dip in cold water until cool enough to handle and cut off kernels to about two-thirds of their depth. Don't scrape the cobs. In a kettle combine chopped red and green peppers, celery, onions, sugar, vinegar, salt, and celery seed; cover until mixture starts to boil; then boil uncovered for 5 min. Blend mustard and turmeric with a little liquid from the kettle, and then add to the mixture. Add corn. Bring to a boil and cook for 5 min. Pour into clean, hot pint jars to 1/2 in. of top. Seal. Process in boiling water bath for 15 min.

If you wish, this relish can be thickened with 1/4 cup flour blended with 1/2 cup water. Add to mixture with the mustard and turmeric. During final cooking, stir constantly.

COWPEAS

Cowpeas are also known as blackeye peas, southern peas, table peas, and field peas. They are great favorites in the South and make equally good eating in the North.

DRYING

Let the beans hang on the plants until the pods and seeds are completely dry. Then shell and package. Or for a somewhat better product, pick the peas when they are green and plump. Shell and blanch seeds in boiling water for 3 min. Then drain well, place in trays in shallow layers, and dry in a 150° oven until the seeds are brittle—about 6 to 10 hr. Package in polyethylene bags or rigid containers and store in a dry, dark, cool place.

FREEZING

Harvest cowpeas when seeds are green and tender. Shell. Then follow standard freezing procedure. Blanch for 2 min. Store for 10–12 mo. To serve, boil for 10–12 min.

CANNING

Raw pack. Harvest when seeds are green and tender. Shell, wash, and pack loosely in pint jars to 1 1/2 in. of top; in quarts to 2 in. of top. Add 1/2 tsp. salt to pints; 1 tsp. to quarts. Cover with boiling water. Leave 1/2 in. head space. Seal. Process in pressure canner at 10 lb. pressure. Pints for 35 min.; quarts for 40 min.

Hot pack. Shell, wash, cover with boiling water, and bring to a boil. Pack loosely in pint jars to 1 1/4 in. of top; in quarts to 1 1/2 in. of top. Add 1/2 tsp. salt to pints; 1 tsp. to quarts. Cover with cooking liquid to 1/2 in. of top. Seal. Process in pressure canner at 10 lb. pressure. Pints for 35 min.; quarts for 40 min.

CRABAPPLES

JELLY

Without added pectin
Makes 5 8-oz. glasses

3 lb. crabapples
3 cups water
4 cups sugar

The redder the skins of the crabapples, the more colorful the jelly. Wash, remove stem and blossom ends, and cut fruit into small pieces. Don't pare or core. Place in a kettle with water, bring to a boil, and simmer till crabapples are soft—about 20 min. Extract juice and measure 4 cups into kettle. Add sugar. Follow standard procedure for making jelly without added pectin.

JELLY

With added pectin
Makes 11 8-oz. glasses

4 lb. crabapples
6 1/2 cups water
7 1/2 cups sugar
1/2 bottle liquid pectin

Wash fruit, remove blossom and stem ends, and cut fruit into small pieces. Don't pare or core. Place in kettle with water, bring to a boil, and simmer for 10 min. Then mash and simmer 5 min. more. Extract juice. Measure 5 cups into kettle and combine with sugar. Follow standard procedure for making jelly with liquid pectin.

The pulp left from the extraction process can be used to make crabapple butter similar to apple butter (see *Apples*).

PICKLED CRABAPPLES

Makes 6 pints

4 lb. crabapples
3 cups cider vinegar
3 cups water
1 tbs. whole allspice
6 cups sugar
1 tbs. stick cinnamon
1 tbs. whole cloves
1 tbs. ground mace

Use bright red crabapples. Combine everything except the fruit and boil till sirup coats a spoon. Wash fruit but don't prepare them otherwise. Add to sirup and heat slowly so that skins will not burst. Simmer till crabapples are tender. Then pack them into hot, sterilized jars, cover with sirup, and seal.

CRABS

In 1971, when we first noticed a lot of people fishing off the bridge over the Lieutenant River, a local tributary of the Connecticut, we thought nothing of it. People were always fishing off the bridge—usually one or two but sometimes, on Sundays, about a dozen. But as the "crowds" continued, we grew curious; and one day we found the answer. One of the men on the bridge was putting together a large wire trap. Crabs!

"Interesting," we said, and thought nothing more about it. Why? we can't tell you. We're crazy about crabmeat. As a matter of fact, only a few weeks earlier we had been on the verge of negotiating an air-express swap of lobsters for crabs with Cousin Carmichael in Mississippi. Mike loves lobster (and crab) but has nothing but crab. We love crab (and lobster) but up to that point had nothing but lobster.

Anyway, we ignored the crabbers. Then, perhaps a week later, a story appeared in one of the newspapers reporting that for the first time in 10 years blue crabs were running in the river. Not just running—they were everywhere. No one knew why they had disappeared a decade earlier. No one knew why they had suddenly returned. But here they were. Even the old-timers were hard put to remember when there had been so many.

"Very, very interesting," we said. And still we did nothing. (The aberrations of the human mind can be strange indeed.)

But finally our indifference disappeared. We were poking along the shores of low, marshy Goose Island in our canoe when we began to spot crabs clinging to the island banks just below the low-water mark. Having nothing to catch them with, we used our hands, and by fluke luck (crabs are awfully fast) we managed to snare three or four.

That did it. When Charlie, our No. 3 son-in-law, came down from Boston the next weekend, we went crabbing in earnest and caught a mess. Such a feast we had!

Moral: When Nature thrusts her bounty upon you, don't ignore it.

(Poor Mike, he's still waiting for his lobster.)

FREEZING

Wash crabs (blue or otherwise) and drop them into boiling salted water for 20 min. Chill in cool salted water until you can handle them; then pick meat from claws and back. Chill as rapidly as possible in a bowl placed in cold water or in the refrigerator. Pack in rigid plastic containers. Freeze. Store for 3 mo. Ready to serve when thawed.

CANNING

Cook crabs, and as soon as they are cool enough to handle, pick them. Dip meat for 1 min. in 1 gal. water to which 1 cup salt has been added. Then dip briefly in 1 gal. water to which 1 cup vinegar has been added. Press out excess moisture. Pack 3/4 cup meat in each half-pint Mason jar. If packing large Pacific crabs, arrange the leg meat on the bottom, top and sides and the body meat in the center. Seal. Process in pressure canner at 5 lb. pressure for 80 min.

CRANBERRIES

DRY STORAGE

Cranberries stored in the coldest part of the refrigerator (at 36° to 40°) will hold for 1–3 mo. They must be kept covered, however, to prevent desiccation.

FREEZING

Wash, stem, and pack cranberries in rigid containers. Don't add sugar. Freeze. Store for 10–12 mo.

CRANBERRY SAUCE

Makes 3 pints

1 lb. cranberries
2 cups sugar
1 1/2 cups water

Sort, wash, and stem berries. Boil with sugar and water for 10 min. Skim. Pour into hot, sterilized jars and seal.

JELLY

Makes 4 8-oz. glasses

1 lb. cranberries
1 3/4 cups water
2 cups sugar

Sort, wash, and stem berries and boil in water until skins crack. Put through sieve or food mill. Combine with sugar and boil just short of jelly stage. Ladle into hot, sterilized glasses and seal.

JUICE

Wash berries. Combine equal measures of berries and water and boil till skins crack. Squeeze juice through jelly bag and strain. Add sugar to taste. Then follow standard juice-making procedure.

CRAYFISH

The crayfish, or crawfish, is a small, fresh-water crustacean very similar to the lobster. It can be preserved in the same ways. The only thing you should do differently is to cook the crayfish in boiling water for 10 min. before cooling and picking. The tails are the only part big enough to be worth eating; the claws may contain a few morsels if they are large, however.

CREVALLE

Also called common jacks, crevalles are angry-looking yellow fish caught mainly off the Florida coast. Wrap in aluminum foil, freeze, and store for 6–9 mo.

CROAKERS

Croakers are rather small, noisy, salt-water fish with lean meat. Freeze in the usual way. Store for 6–9 mo. The fish can also be cold-smoked or salted.

CUCUMBERS

There are two general types of cucumbers—slicing and pickling. Most home gardeners grow the slicing type because they are somewhat better to eat fresh and they make perfectly good pickles. However, pickling cucumbers have the edge with pickle packers because they are short and blocky, and the plants are more productive.

BRINING

See Chapter 6.

FREEZING

You can freeze them, if you want to eat chilled cucumber slices in winter, but you won't be exactly overwhelmed by the end product. Just pare; cut very thin slices directly into container, seal, and freeze. Store for 10–12 mo.

DILL PICKLES

Makes 9–10 quarts

20 lb. cucumbers about 4 in. long
3/4 cup mixed pickling spices
30 heads green or dry dill weed or
10 tbs. dill seed
1 3/4 cups salt

2 1/2 cups vinegar
2 1/2 gal. water

Place half the spice and a layer of dill in a 5-gal. crock. Fill in within 3 in. of top with washed cucumbers. (Put these in carefully; don't drop them lest you bruise them.) Cover with half the spice and a layer of dill. Mix salt, vinegar, and water and pour over the cucumbers. Let the cucumbers ferment according to directions in Chapter 13. When the cucumbers turn olive green, are free of white spots, and are translucent inside, they are ready for packing.

Pack into clean, hot quart Mason jars, but don't wedge too tight. Add several pieces of the dill from the crock. Strain the brine through cheesecloth, heat to boiling, and pour it over the cucumbers to 1/2 in. of top of jars. Seal. Process in boiling water for 15 min. Start timing as soon as jars are placed into boiling water.

SWEET GHERKINS

Makes 7–8 pints

5 lb. cucumbers less than 3 in. long
1/2 cup salt
8 cups sugar
6 cups vinegar
3/4 tsp. turmeric
2 tsp. celery seed
2 tsp. mixed pickling spices
8 1-in. pieces stick cinnamon
1/2 tsp. fennel

This pickle takes 4 days to make. Cucumber stems need not be removed. Wash cucumbers, place in a kettle, and cover with boiling water. About 6 hr. later, drain and cover with fresh boiling water. Allow to stand overnight and repeat process. Once again, 6 hr. later, drain and cover with fresh boiling water to which the salt has been added. Let stand overnight.

Drain. Prick cucumbers 3 times with a sharp table fork. Mix 3 cups sugar and 3 cups vinegar, add spices, bring to a boil and pour over cucumbers. Six hours later, drain sirup into a kettle, add 2 cups sugar and 2 cups vinegar, bring to a boil, and pour over cucumbers again. Let stand overnight.

In the morning, drain sirup into kettle, add 2 cups sugar and 1 cup vinegar, bring to a boil, and pour over pickles. Six hours later, drain off sirup once more, add 1 last cup sugar, and bring to a boil. Pack cucumbers into clean, hot jars and cover with sirup to 1/2 in. of top. Seal. Place jars in boiling water bath and process for 5 min. after water comes to a boil.

SOUR PICKLES

Makes 8 pints

32 4-in. cucumbers
1 cup salt
4 qt. water
4 cups vinegar
1 cup sugar
1/2 tbs. whole cloves
1/2 tbs. mustard seed
1/2 tbs. celery seed
1/2 tbs. peppercorns

Wash cucumbers, place in a crock, and cover for 24 hr. with brine made of the salt and water. Drain and cover with fresh water for 20 min. Drain. If cucumbers are still very salty, soak in fresh water once more. Then cut cucumbers lengthwise, either in halves or strips, depending on the size, and put them in cleaned crock. Boil vinegar, sugar, and spices (tied in a bag) for 5 min. Remove spice bag and pour sirup over cucumbers. Let stand for 24 hr. Drain sirup into a saucepan and bring to a full boil. Meanwhile pack cucumbers in hot, sterilized jars. Cover the cucumbers with the sirup and seal jars.

CRISP PICKLE SLICES

Makes 7–8 pints

6 lb. medium cucumbers
6 medium white onions
3 garlic cloves
1 green pepper
1 red sweet pepper
1/3 cup salt
5 cups sugar
1 1/2 tsp. turmeric
1 1/2 tsp. celery seed
3 cups white vinegar
2 tbs. mustard seed

Wash and slice unpeeled cucumbers thin. Measure out 4 qt. Slice onions. Chop garlic cloves. Slice peppers in strips, discarding interiors. Mix these ingredients together with the salt in a large kettle, cover with ice cubes, and mix again. Let stand for 3 hr. Drain completely. Add all other ingredients and bring to a rolling boil. Then pour into hot, sterilized jars and seal.

OLIVE OIL PICKLES

Makes 5–6 pints

4 lb. medium cucumbers
2 lb. large white onions
1/2 cup salt
3 red chile peppers
1 qt. vinegar
3 cups brown sugar
1 tbs. turmeric
1 tbs. celery seed
1 tbs. mustard seed
3 tbs. olive oil

This is the pickle that floored daughter Cary. Wash and peel cucumbers and slice thin. Cut onions into 1/4-in. slices. Dissolve salt in 2 qt. water, cover vegetables, and let them stand overnight. In the morning drain completely. Slice peppers into strips. Combine with vinegar, sugar, and spices and bring to a rolling boil. Add cucumbers and onions, bring to a boil, and boil for 1 min. Drain off the liquid into another kettle, bring to a boil, and add olive oil. Pack cucumbers and onions into hot, sterilized jars. Cover with the liquid. Seal. Let pickles stand for 3 or 4 days before using.

BREAD AND BUTTER PICKLES

Makes 7–8 pints

8 lb. cucumbers
2 lb. white onions
1/2 cup salt
1 qt. vinegar
2 cups sugar
2 tsp. celery seed
2 tsp. turmeric
1 tsp. dry mustard

Cut unpeeled cucumbers and peeled onions in slices about 1/8 in. thick. Arrange in layers in a large kettle, sprinkling each layer with salt. Let stand for 3 hr., then drain and rinse. Bring other ingredients to a boil and add vegetables. Bring to a boil again and cook for 5 min. Pack into hot, sterilized jars and seal.

CUMIN

Cumin seeds are an important ingredient of some pickles and are also used for flavoring. They are produced by a tiny annual plant that can be grown almost everywhere. To harvest the seeds when dry, pull up the entire plant and shake the seeds out on cheesecloth. Pack in bottles.

CURLED CHERVIL

This annual herb resembles parsley and tastes like parsley crossed with fennel. The leaves are used in meats, soups, stews, omelets, and salads. They are usually dried and stored in airtight containers; but young, tender leaves can also be frozen without blanching. In the latter case, pack the leaves in very small quantities so that you don't have to thaw a lot to use a few.

CURRANTS

FREEZING

Wash and stem berries. If packing them whole, you can freeze them without sweetening or with 1 lb. sugar to every 2 or 3 lb. fruit. If packing crushed berries, mix 1 lb. sugar with 2 to 3 lb. crushed fruit. Store for 10–12 mo.

DRYING

Wash and stem berries. Spread in cheesecloth-lined trays not more than 2 berries deep. Dry in oven until berries rattle and do not leave moisture on your fingers when you crush them. Start at 135° and increase gradually to 150°. Store in airtight containers in a cool, dark place.

JELLY

Without added pectin
Makes 4 8-oz. glasses

4 lb. currants
1 cup water
4 cups sugar

Wash and crush currants, but don't bother about stems. Combine with water, bring to a boil, then simmer for 10 min. Extract juice and measure 4 cups into a kettle. Add sugar. Follow standard procedure for making jelly without added pectin.

JELLY

With added pectin
Makes 10 8-oz. glasses

4 lb. currants
1 cup water
7 cups sugar
1/2 bottle liquid pectin

Wash, crush, but do not stem currants. Bring to boil in water and simmer for 10 min. Extract juice. Measure 5 cups into a kettle with sugar. Follow standard procedure for making jelly with liquid pectin.

BAR-LE-DUC

Makes 5 8-oz. glasses

4 lb. currants
7 cups sugar
1/4 cup water

Wash, crush, but do not stem 1 lb. currants. Add water, bring to a boil, and simmer for 10 min. Extract 1 cup juice. Wash and stem remaining currants and combine with juice and 4 cups sugar. Cook on high heat, stirring constantly, for 5 min. Then remove from burner and let stand overnight. The next day, add remaining sugar, bring to a boil, and cook to jelly stage—about 30 min. Stir frequently. Skim. Ladle into hot, sterilized glasses and seal.

CUSK

The cusk is a lean-meated relative of the cod and caught in the same North Atlantic waters. You can freeze and store it for 7–10 mo. You can also smoke or salt it.

DANDELIONS

Dandelions to be eaten as greens (or in salads) must be picked while they are young, before flower buds appear; otherwise they will be very bitter. Wash thoroughly. Then freeze or can like other greens.

WINE

You must pick a gallon of dandelion flowers for each gallon of wine. Use new flowers and remove all green parts from the heads. Place in primary fermenter and cover with a like quantity of boiling water. Let the flowers steep for 4 days, stirring daily. Then strain the liquid through cheesecloth into a large kettle. For each gallon add 1 tsp. nutmeg, 1/2 tsp. cloves, 3 oranges, and 1 lemon cut in 1/2-in. slices. Boil for 30 min., cool, and return to the primary fermenter. Mix in 3 lb. sugar and 1 package dried wine yeast. Cover and let ferment for a week. Remove fruit and continue to ferment for another week or so. When action subsides, rack into another fermenter, which need not be covered, and allow fermentation to continue until it stops completely. Clarify wine as necessary, and bottle. The wine should not be drunk until it has aged 6 mo.

DASHEEN

Dasheen is a relative of the elephant ear and, like it, grows in warm climates. In the tropics it is called taro. It produces small tubers, which are eaten like sweet potatoes. Dig these up before the first fall frost and store them in a dark, well-ventilated, humid place at 50°.

DATES

Dates grow in enormous clusters on the handsome, upright, spreading date palm. The trees are grown on a wide scale only in the desert valleys of Arizona and Southern California; but isolated trees which sometimes bear fruit are found in other warm areas.

Dates are ready for harvesting from September through November. Pick the fruits when they become translucent—that is, when the flesh becomes soft, pliable, and light brown. (The fruits within a cluster do not all reach this stage at the same time.)

Pick over the fruits carefully, discarding those that have damaged skins, are covered with a fungus, or smell sour. Wipe the remainder with a damp turkish towel. Never wash dates.

To mature the dates and destroy any insects or insect eggs which may be on them, spread them in a single layer on a metal rack, aluminum screen, or perforated aluminum sheet. A cookie sheet may also be used, but it does not allow the heat to circulate around the fruit as well.

If possible, buy a laboratory thermometer and insert it in one of the dates. Turn on the oven until the thermometer reaches a temperature of 120° to 125°; then turn the oven off until the temperature drops to 105°–110°. Then turn the oven on again. Repeat this process until the dates are ready for storage. (This may take several days.)

If you don't have a thermometer, another but less reliable way to manage the oven is to preheat it to 200° and turn it off. Slide in the dates and leave them there until they are partially cool. Then remove them, turn on the heat, and repeat the process again and again.

If your dates are semidry varieties (the most common are Deglet Noor, Zahedi, and Medjool), they are done when you can squeeze the fruits to the seed without breaking the skin or pulp. The fruits are soft but not wet. They can then be packed in semiairtight containers and stored for a year or more. Deglet Noor and Zahedi are stored in the coldest part of the refrigerator fresh-food compartment. Medjool should go into a freezer.

If you want to dry these varieties further, heat the fruits in the oven until firm and chewy. Then pack them in airtight containers such as plastic freezer boxes or Mason jars, and store in a dark place at normal room temperature. They will keep for several years.

Soft varieties of dates, such as Kadrawi, Barhee, and Hayany, can be dehydrated in the same way to any degree of dryness. But they are normally dried only until they reach a point where they are wet inside and very soft. They are easily injured when handled. These should be packed in rigid plastic containers and stored in the freezer. They can be held for over a year, but it is better to use them within 6 months because their flavor deteriorates with long storage.

It is possible to take dates directly from the tree and freeze them. But as R. H. Hilgeman, horticulturist in the University of Arizona's College of Agriculture, points out, "Curing in the oven for at least a short time improves flavor and breaks down part of the fiber so that eating qualities are improved."

PICKLED DATES

Slit fresh dates lengthwise, remove pits, and insert halves of pecan meats. Pack in clean jars. Meanwhile make a sirup in the proportions of 8 cups sugar, 3 cups white wine vinegar, 1 cup water, 1 tsp. oil of cinnamon, and 1/2 tsp. oil of cloves. Boil this for 3 min., pour over dates to 1/2 in. of top of jar. Seal. Process in pressure canner at 5 lb. pressure for 10 min.

DATE PASTE

Clean fresh dates, remove pits, and put in top of a double boiler over boiling water. Cook for 30 min., stirring often. Then pour the thick mixture into hot, clean Mason jars. Seal. Process in a boiling water bath for 10 min.

DEWBERRIES

Dewberries are blackberries and are put up in the same ways.

DILL

Dill is an attractive annual plant with feathery foliage which is used to flavor fish, stews, salads, sauces, and potatoes. Leaves to be dried are collected before the flowers open. A few sprigs cut at the same time can also be added to a bottle of vinegar to give it a special flavor.

If you prefer fresh dill weed to dried, spread freshly cut leaves on a cookie sheet or aluminum foil and put in the freezer until hard. Then package in polyethylene bags or rigid containers and return to freezer. Store for 10–12 mo.

Dill seeds are collected when dry, and bottled and used in making pickles.

DOVES

Draw doves as soon as possible after shooting. Dry-pluck. Cut off feet, outer wing bones, and neck. Let age for 24 hr. or more in the refrigerator. Then freeze. Store for 9 mo.

DRUM

The black and red drums are lean-meated, salt-water fish caught from Long Island south. The former reaches 150 lb.; the latter is much lighter but considered more fun to catch. Both are preserved by freezing in aluminum foil. Store for 6–9 mo.

DUCK

FREEZING

See Chapter 14 for how to kill, pluck, and clean poultry. Let the ducks chill thoroughly in the refrigerator. Then wrap whole birds in aluminum foil or polyethylene bags and freeze. Store for 4–6 mo.

CANNING

Pluck, clean, and chill ducks. Then follow standard canning procedure for meat and poultry. Cut birds into pieces and process by either the raw-pack or hot-pack methods. If bone is removed, process pints for 75 min.; quarts for 90 min. If bone is left in, process pints for 65 min.; quarts for 75 min.

DUCK, WILD

Some hunters believe that wild duck is improved by bleeding. If you concur, slit the throat after picking up a bird and let it hang. In any case, put each bird by itself where it can cool rapidly. Don't throw all birds together, because this retards cooling.

Dress ducks as soon as possible by cutting a slit from the breastbone down to and around the vent. Remove entrails and wipe out the cavity with a damp cloth or cleansing tissue. Make a cut near the base of the neck and remove the crop.

You can now let the birds hang in a cool place for 2 or 3 days so as to age and tenderize the meat; or you can pluck them immediately and let them age for the same length of time in the refrigerator.

Cut off the wings and the feet at the first joint. Dry-pluck the birds. Skinning is easier, but the birds lose moisture and flavor.

Wrap each duck in aluminum foil, taking pains to work out as much air as possible. Or put the bird in a polyethylene bag. Freeze. Store for 4–6 mo.

EEL

Americans have never developed the Scandanavians' appreciation of fresh-water eels. This may be partly because eels have a wretched way of turning up in likely looking fishing holes and wrecking efforts to catch anything else.

We remember a hole in the stream below the outlet of Lower Shin Pond, in northern Maine, which always looked in summer as if it should be teeming with trout. But every time we tried our luck—and we did so quite a few times—we caught nothing but eels. They were everywhere, and all very willing to swallow hook and line deep down inside them.

But eels are good eating; and if you can't catch anything else, you might as well be happy with them.

FREEZING

Behead the fish and strip off the skin back to the tail the way you peel off a stocking. Remove entrails and wash clean. Cut into 3-in. pieces. Package in polyethylene bags and freeze. Store for 6–9 mo.

SMOKING

Prepare the eels as above or leave the heads on. After brining and drying, put in the smokehouse. Cut chunks of eel are spread on wire mesh. String wires through the heads of whole eels and hang them. Hot-smoke according to directions in Chapter 4.

PICKLED EEL

Cut cleaned, skinned eels into pieces about 5/8 in. thick. Wash and drain; then dredge with salt and let stand for 1 hr. Rinse off salt, dry, and rub with garlic. Brush with vegetable oil and broil in the oven until light brown. Allow to cool on paper towels.

For 10 lb. fish, mix together 6 sliced, medium onions; 18 bay leaves; 9 tsp. whole allspice; 5 tsp. peppercorns; 4 tsp. whole cloves; 4 tsp. mustard seed; 3 tsp. ground mace. Place the pieces of eel in layers in a crock and sprinkle seasonings between them. Weight the fish down, cover tightly, and store in the refrigerator for 24 hr. Then mix 3 parts white vinegar with 1 part water and pour over the fish. Cover and let stand in the refrigerator for 48 hr. before using. For storage, the eel can be left in the crock or repacked with the spices in sterilized jars. Cover with the vinegar, seal, and store in the refrigerator.

EGGPLANT

FREEZING

Peel and cut mature eggplants into 1/3-in. slices. To prevent darkening before cooking (though some darkening will occur during cooking anyway), put the slices as you cut them into a solution of 2 tbs. salt and 2 qt. water. Follow standard freezing procedure. Blanch for 3 min. To make it easier to separate slices when you use them, place pieces of aluminum foil or freezer wrap between them. Overwrap the bundle of slices with aluminum foil. Store for 10–12 mo. To serve, sauté in oil.

RATATOUILLE

In this book it has been our policy not to give many recipes for prepared dishes that can be frozen; but this is such a good one we violate our rule. It is our favorite way to preserve excess quantities of eggplant, zucchini, and tomatoes.

1 eggplant
1 large zucchini
6 tomatoes
2 onions
4 tbs. olive oil
1 clove garlic
1 tbs. chopped parsley
1 bay leaf
salt and pepper
Worcestershire sauce

Peel eggplant. Cut eggplant and zucchini in 1-in. or somewhat larger cubes and boil for 10 min. Drain. Peel tomatoes and cut into pieces. Chop onion fine and brown in oil. Chop garlic and add with all other ingredients to onion. Simmer till tender, about 20 min. Allow to cool thoroughly. Then seal in rigid containers and freeze. To serve, heat in a double boiler. Store for 10–12 mo.

EGGS

The ideal storage conditions for eggs require a temperature of 29° and a relative humidity of about 82 percent. Seventy percent of eggs stored under these conditions will have a grade A rating at the end of 6 mo.

Unfortunately, such conditions are next to impossible to maintain in the home. The best you can do is to put the eggs in the meat compartment of a refrigerator. Make sure the eggs are very fresh. Don't wash them. If the shells are thin, dip them in mineral oil. The majority of eggs stored in this way should keep for about 3 mo.

FREEZING

Whole eggs. Break eggs into a bowl and mix lightly without beating in air. For baking, mix sugar or corn sirup into the eggs at the rate of 1 tbs. per 1 cup egg. For scrambling, souffles, etc., mix 1 tsp. salt with 1 cup eggs. Pour into rigid plastic containers and freeze.

After mixing with salt, sugar, or sirup, eggs can be ladled into muffin pans in one-egg portions and frozen. They should then be turned out of the pans, wrapped individually in foil or small polyethylene bags and returned to the freezer.

Store whole eggs for 8 mo. To use, thaw in refrigerator.

Egg yolks. Mix with sugar, sirup, or salt in the above proportions; pour into rigid containers and freeze. Store for 8 mo. Thaw in refrigerator.

Egg whites. Pour into rigid containers and freeze. Store for 10 mo. Thaw in refrigerator.

ELDERBERRIES

Although elderberries grow pretty much throughout the United States—in the wild and under cultivation—they are not widely known. But the small, sweet, blue, purple, or red fruits, which are borne in huge clusters, make very good jam, jelly, and pies.

FREEZING

Wash and stem elderberries and pack in rigid containers without sweetening. Seal and freeze. If skins of berries seem a bit tough, place them in boiling water for 1 min. before packaging. Store for 10–12 mo.

CANNING

Wash and stem berries. Mix 1/4 cup sugar with each 4 cups berries and let stand for 2 hr. Then cook until berries are hot and sugar dissolves. Pack in jars to 1/2 in. of top. Add boiling water if there is not enough sirup to cover berries. Leave 1/2 in. head space. Seal. Process in boiling water. Pints for 10 min.; quarts for 15 min.

JELLY

Makes 10 8-oz. glasses

3 lb. elderberries
1/2 cup lemon juice
7 cups sugar
1 bottle liquid pectin

Wash, stem, and crush berries. Cook gently until juice flows then simmer for 15 min. Extract 3 cups juice and combine in a kettle with lemon juice and sugar. Follow standard procedure for making jelly with liquid pectin.

JAM

Makes 6 8-oz. glasses

3 lb. elderberries
6 cups sugar
1/2 cup lemon juice

Combine and bring all ingredients to a boil, stirring to prevent sticking. Follow standard procedure for making jam without added pectin.

WINE

Pick over ripe elderberries; cut out thicker stems, but don't bother with all the thin ones that lace the berries close together. Place in a primary fermenter and cover with water. Stir in 1 lb. sugar per gallon of the mixture and add one-fifth package dry wine yeast for each gallon. Then follow directions in Chapter 16. Do not add any more sugar unless the juice is sour.

ELK

Elk is handled and preserved like venison. If your elk is killed when the weather is mild, it is advisable to skin it at once so the carcass can cool rapidly.

ENDIVE

Also called escarole, endive is a leafy vegetable very similar to lettuce and used in the same way. It is best grown as a fall crop. You can save some of the heads for use late in the fall by taking up the plants with a large ball of earth before the first hard frost and putting them in a coldframe or in boxes in a bright, cool basement. They should not be allowed to freeze or dry out. Before using, blanch the hearts by drawing up the outer leaves and tying them together loosely over the hearts.

FENNEL

Also called Florence fennel and finocchio, this is an attractive, feathery annual with an anise flavor. Dry leaves in a cool, airy place and seal in small containers. To put up seeds, which are used in pickles, for flavoring vegetables, fish, meats, and cheese and egg dishes and also in baking, harvest flower heads when dry and pack the seeds in bottles.

FIDDLEHEADS

Fiddleheads are the young, unopened fronds of ferns which are eaten raw in salads and cooked like greens. The name stems from the fact that the tightly coiled top of the frond resembles the head of a violin. Fiddleheads are also called crosiers because of their resemblance to a bishop's pastoral staff.

While the fiddleheads produced in early spring by most species of deciduous ferns are edible, the best are those from brakes *(Pteridium aquilinum)*, male ferns *(Dryopteris filix-mas)*, and ostrich ferns *(Pteretis nodulosa)*.

Collect fiddleheads when they are about 6 in. high and tender enough to be snapped off with your fingers. Don't remove more than a third to a half of the shoots from any one plant. Preserve by standard freezing procedure. Before blanching, pull each stalk through your fingers to remove the feltlike material covering it, and wash briefly in cold water. Blanch for 2 min. Store for 10–12 mo. To serve, cook for 5 min.

Fiddleheads can also be canned. See *Greens*.

FIGS

FREEZING

Use ripe figs that are not sour in the middle. Wash, cut off stems, and slice or leave fruit whole. Pack in rigid containers. Cover with 35 percent sugar sirup to which ascorbic acid has been added. (Use 4 tsp. ascorbic acid to 4 cups of sirup.)

You can also freeze figs without sweetening. Pack dark-colored figs dry. Cover light-colored figs with water to which ascorbic acid has been added in the above proportions.

Storage life in all cases is 10–12 mo.

DRYING

Use fully ripe but partly dried figs. Wash, stem, and cut in halves or leave whole. Peel if desired. Mix 1 cup sugar with 3 cups water, bring to a boil, and add figs. Simmer for 10 min., then remove from heat and let stand for 10 min. more. Drain. Spread in trays in a single layer and dry in oven till leathery and with pliable flesh but a little sticky. Start at 115° and gradually increase to 145°.

Instead of sirup-blanching figs, you may blanch cut figs in steam for 20 min.; or simply dip whole figs in a brine made with 5 tsp. salt and 1 qt. water; or sulfur Kadota figs for 2 to 3 hr. The end product, in these cases, is a little less sweet.

CANDIED FIGS

See Chapter 10. The figs are left whole.

JAM

Without added pectin
Makes 10 8-oz. glasses

5 lb. figs
3/4 cup water
6 cups sugar

Cover washed figs with boiling water and let stand for 10 min. Then drain and remove stems. Chop fruit. Measure 8 cups into a kettle, add water and sugar. Follow standard procedure for making jam without added pectin.

JAM

With added pectin
Makes 11 8-oz. glasses

3 lb. figs
1/2 cup lemon juice
7 1/2 cups sugar
1/2 bottle liquid pectin

Wash, remove stems, and grind figs. Measure 4 cups into a kettle, add lemon juice and sugar. Follow standard procedure for making jam with liquid pectin.

PRESERVES

Makes 10 half-pints

4 1/4 lb. figs
2 lemons
1 1/2 qt. water
7 cups sugar
1/4 cup lemon juice

Wash, stem, and peel figs. Slice lemons thin. Heat water, add sugar and lemon juice, and cook till sugar dissolves. Add figs and boil for 10 min., stirring occasionally. Add sliced lemons and continue boiling till figs are translucent—about 10 min. (Add a little boiling water to sirup if it gets too thick before figs are clear.) Cover and let stand for 24 hr. Pack in clean jars to 1/4 in. of top. Seal. Process in simmering water to cover for 30 min.

PICKLED FIGS

Makes 7 pints

8 lb. figs
1 tsp. salt
1 gal. water
8 cups brown sugar
1 qt. cider vinegar
7 or 8 pieces stick cinnamon
2 tsp. whole cloves

Wash and stem figs. Boil salt and water; add figs and boil for 15 min. Drain thoroughly. Bring sugar and vinegar to a boil and add spices and figs. Simmer for 1 hr. Pack in hot, sterilized jars, making sure sirup covers the fruit. Seal.

FILBERTS

Filberts, or hazelnuts, often grow on such small shrubs that it is easy to pick the nuts by hand, but the trouble with this method is that the nuts may not be mature. It's far better to let Nature take her course. When the nuts are ready for harvesting, the husks usually open and the nuts fall to the ground. Sometimes, however, the husks and nuts come down together.

Prompt collection of fallen nuts is not so essential as it is with many other nuts; nevertheless, you shouldn't let them lie around too long because they will gradually discolor.

Once harvested, drop the nuts in a tub of water to sort out the bad, which float, from the good, which sink. Then spread them in trays only a few layers deep and let them dry in a warm, dry, well-ventilated place. Stir frequently. Or you can hasten the curing process by heating the air around them to about 90° with an electric fan-heater. Drying is complete if a cold kernel snaps when bitten.

Store the nuts in sacks in a cool, dry place. Kernels can also be frozen or canned.

FLOUNDER

If you lead a fast-paced life, the flounder, or flat fish, is one of your best friends, because it can be taken out of the freezer, thawed quickly, and cooked just as quickly in many delightful ways. Having long ago settled down to full-time writing careers, we no longer move at a hectic pace, but we still like to keep a supply of small flounders —whole or filleted—in the freezer.

To freeze, just clean; wrap in aluminum foil or slip into polyethylene bags; and put in the freezer. Store for 6–9 mo.

Flounder can also be given a cold or hot smoke. Cold is better.

FLUKE

Flukes are flat fishes related to flounders and handled in the same way.

FROGS

If you are city born and bred, the sounds of the country may be more alarming than appealing when you first hear them. Many will certainly be more annoying, and that is particularly true of the bullfrogs' "chugarum, chugarum." If you camp at night near a pond well stocked with frogs, the incessant rumbling of the basses, baritones, and tenors may keep you awake till dawn.

Once you get used to it, however, frog talk becomes a lullaby. It has the same haunting, reassuring qualities as the foghorns in San Francisco Bay. You listen for a few minutes and then you're asleep.

The frogs talk on.

But there is one thing about frogs even better than their chatter: their legs. Most people compare the meat to chicken, but it is far superior to any chicken (provided it isn't overpowered with a Frenchman's garlic). If you fry the legs briefly in cornmeal or breadcrumbs, you can eat a dozen pair without batting an eye.

Hunting for frogs isn't much of a sport because they sit out in plain sight in the sunshine, and they rarely move until you are almost close enough to touch them. The most humane way to kill them is to shoot them in the head with a .22. Another somewhat less humane way is to spear them with a three-tined, forklike spear on a long pole. The least humane way, but probably the surest, is to tie three large fishhooks together to form a three-pronged star, bait them with a piece of red flannel, and hang them from a stick on a 4-ft. piece of fishing line. Most frogs jump at this lure, mouth open, if you drop it in front of them. If they don't, you can hook them under the chin. In either case, just haul them into your boat and kill them with one blow of a heavy club.

Butcher frogs by cutting the hind legs off at the hip. You can then peel the skin off like a nylon stocking. Discard the rest of the body; it probably contains some meat, but not very much.

Package frogs' legs in polyethylene bags

or rigid plastic containers and put them in your freezer. Store for 6 mo.

GALLINULE

The gallinule is a ducklike bird which is hunted to a limited extent in the East and along our southern border. For how to handle and preserve it, see *Duck, wild.*

GARLIC

Pull up the bulbs when the tops fall over and let them dry, without washing, in a shady place. Store in a dry, cold, well-ventilated place. The bulbs will keep for up to 8 mo.

GINGER

The fleshy roots of any species of ginger plant can be used to make the spice called ginger, but the most common source of the spice is *Zingiber officinale,* a 3-ft. perennial which grows in our warmest climates.

Use only succulent roots. After digging them up, allow them to dry in an airy place out of the direct sun for about 2 days. The roots can then be stored for several weeks at 55° in a humid place and used fresh. Or you can continue drying the roots until completely dry, then scrape off the skin, and grind fine for storage in tight containers. Freshly dug roots can also be wrapped in aluminum foil and frozen.

CANDIED GINGER

Use young, green ginger roots if you can lay your hands on them. Older roots are fibrous and hot. Scrape the roots to remove the skin, and cut them into 1/8-in. pieces. Cover with water, boil for 5 min., and drain. Repeat this process 3 more times. After the last boiling, however, save the liquid and measure the combined ginger and water. To this add 1 1/2 times as much sugar, and boil till ginger is clear.

If you like dry, crystallized ginger, drain off the sirup as completely as possible and roll the ginger in granulated sugar. Pack in tight glass jars. If you prefer a sirupy product, pour the ginger and sirup directly into hot, sterilized jars and seal at once.

GOAT

Young kid is a favorite food of many Americans of foreign extraction. It can be preserved in several ways like lamb, but is best when frozen.

Goat's milk is used to make many superb cheeses, but, unfortunately, their manufacture is too complicated for anyone without a factory.

GOOD KING HENRY

Also called mercury, this is a tall, vigorous perennial with triangular leaves which may be eaten as greens. For how to preserve, see *Greens.*

GOOSE

Handle domestic and wild geese in the same way. The latter, however, should be allowed to age in a cool place for 2 or 3 days before processing. Pick birds dry if you can.

FREEZING

Wrap prepared birds in aluminum foil or large polyethylene bags and put in the freezer. Store for 4–6 mo.

CANNING

Follow standard procedure for canning meat and poultry. Cut goose into pieces and remove excess fat (of which geese have plenty). Process by either hot-pack or raw-pack method. If birds are boned, process pints for 75 min.; quarts for 90 min. If birds are not boned, process pints for 65 min.; quarts for 75 min.

GOOSEBERRIES

FREEZING

Wash fruit, remove blossom, and stem ends, drain and pack dry in rigid containers. Store for 10–12 mo.

CANNING

Make a 40 percent sugar sirup. Wash and stem berries and pack in jars to 1/2 in. of top. Shake the berries down but don't pack tightly. Cover with boiling sirup to 1/2 in. of top. Seal. Process in boiling water. Pints for 10 min.; quarts for 15 min.

JAM

Without added pectin
Makes 7 8-oz. glasses

4 qt. gooseberries
6 cups sugar

Wash, stem, and crush berries. Measure 9 cups into a kettle. Mix with sugar. Follow standard procedure for making jam without added pectin.

JAM

With added pectin
Makes 9 8-oz. glasses

2 qt. gooseberries
6 cups sugar
1/2 bottle liquid pectin.

Wash, stem, and crush berries. Measure 4 cups into a kettle and add sugar. Follow standard procedure for making jam with liquid pectin.

GRAPEFRUIT

Grapefruit like other citrus fruits can be stored on the tree or in boxes in a cool storage room for a considerable period of time. See comment under *Oranges.*

FREEZING

Peel and section fruit. Pack in rigid containers and cover with 30 percent sugar sirup. The sirup may be made in part of excess fruit juice. Store for 6 mo.

CANNING

Grapefruit is better canned in tin than in glass, but use of glass jars is quite permissible. Peel and section fruit. Be sure to remove white membranes. Pack segments firmly in hot jars to 1/2 in. of top. Add 3 tbs. 40 percent sugar sirup to pints; 6 tbs. to quarts. Seal. Process in simmering water to cover. Pints for 10 min.; quarts for 15 min.

JUICE

Extract juice from fruit with a reamer, not a press. If juice is to be canned, follow standard procedure described in Chapter 9. If juice is to be frozen, pour into glass freezer jars (plastic may affect flavor), and freeze. If desired, sweeten juice with 2 tbs. sugar per quart. Store for 3–4 mo.

MARMALADE

Carefully peel 1 or 2 grapefruit, lay peel flat, and pare off and discard about half of the white part. Boil for 10 min. in water; drain; and then repeat boiling and draining twice more. Slice peel into thin slivers. Chop pulp and remove seeds and rag. Add to sliced peel. Cover with 1 qt. water and let stand for 12 hr. or more. Then boil till peel is tender—about 40 min.

Measure fruit mixture, and for each cup add 1 cup sugar. Bring to a boil slowly, stirring, and boil hard until just short of jelly stage. Stir frequently. Ladle into hot, sterilized glasses and seal. If you use 4 cups sugar, this recipe will make 3 8-oz. glasses.

JELLY

Makes 5 8-oz. glasses

4 medium grapefruit
2 qt. water
3 cups sugar
yellow food coloring

Peel grapefruit and cut in small pieces. Pour off juice and weigh it, then add enough pulp to make 2 lb. juice and pulp combined. Combine with water in a large kettle and boil for 5 min. Extract juice and combine it with the sugar and a few drops of food coloring in a kettle. Boil to jelly stage. Ladle into hot, sterilized glasses and seal.

This jelly is too sweetly bitter to suit our tastes, but you may like it. If not, you might try substituting 3 oranges for two of the grapefruit.

CANDIED GRAPEFRUIT PEEL

1 grapefruit
2 cups sugar
1 cup water

Remove peel and cut into narrow, lengthwise strips. Cover with water in a saucepan and bring to a boil. Drain. Repeat process three more times to remove bitterness from peel. Combine sugar and 1 cup water, bring to a boil and add the peel. Simmer till peel starts to look transparent; then remove from heat and let stand for 24 hr. Return to range and boil to 222°. Pack peel in hot, sterilized glasses, cover with sirup and seal. Do not open for at least 6 wk. Peel will keep 3–4 mo.

When you're ready to use the peel, remove it from the glasses and dip it in hot water to remove the sirup. Dry in a 175° oven—with the door open—for 6 to 8 hr. Then heat 1/2 cup water and 3 cups sugar till sugar dissolves. Dip peel in this, place on a wire rack and dry in a 175° oven. Store in jars with lids loose or with several small holes punched in them. The peel will keep for about 3 mo.

GRAPES

FREEZING

Wash and stem grapes. Cut seedy fruits in half and remove seeds. Put in rigid containers and cover with 40 percent sugar sirup. Store for 10–12 mo.

CANNING

Wash and stem grapes and remove seeds. Pack in jars to 1/2 in. of top and cover with 50 percent sugar sirup. Put open jars in warm water to 2 in. of rims and boil for 30 min. Add sirup if necessary. Seal. Process in boiling water. Pints for 10 min.; quarts for 12 min.

RAISINS

Use seedless varieties of grapes. Wash and stem them. Place in trays in a single layer. Dry in the oven till leathery and pliable. The temperature should not exceed 150°. Store in tight containers in a dry, cool, dark place.

JUICE

To extract juice from purple grapes, remove stems and crush fruit. Then heat to 160° and squeeze through a jelly bag. To extract juice from red grapes, remove stems and put a quart of grapes in a cheesecloth bag. Immerse in boiling water for 30 sec. Then crush. Let stand 10 min. Then squeeze through jelly bag. To extract juice from white grapes, remove stems, crush fruit, and squeeze through a jelly bag.

From here on all grapes are handled in the same way. Don't add sugar. Process according to directions in Chapter 9.

SIRUP

1 1/4 cups grape juice
1 1/2 cups sugar
1/4 cup white corn syrup
1 tbs. lemon juice

Combine ingredients in a large kettle, bring to a rolling boil, and boil for 1 min. Remove from heat and skim. Pour into hot, clean Mason jars. Seal. Process in boiling water for 10 min.

GRAPE SHRUB

2 qt. grapes
1 pt. cider vinegar
sugar

Wash and stem grapes, place in a crock or nonmetallic container, and add vinegar. Cover with cheesecloth and let stand for 4 days, stirring daily. Then strain through a jelly bag. Don't squeeze. Add 2 cups sugar to each 2 cups juice. Bring to a boil and simmer for 5 min. Pour into hot, sterilized jars and seal. To serve, dilute with cold water.

WINE

The process for making red grape wine is described in Chapter 16. Until you become an expert vintner, it's advisable not to attempt white wine.

Concord is a favorite variety for making red wine but is actually not so good as many others. Better varieties growing in the East are Eumelan, Steuben, Champanel, Chelois, Chambourcin, Baco Noir, Landal, and Seibel 7053. Excellent western grapes are Cabernet Sauvignon, Zinfandel, Gamay, and Tannat.

JELLY

Without added pectin
Makes 4 8-oz. glasses

3 1/2 lb. grapes
1/2 cup water
3 cups sugar

Concord grapes and similar varieties make the best purple jelly, but other American bunch grapes can be used. We recently used white grapes with just a few purple fruits added, and they made an outstanding jelly in a very lovely shade of rose.

Wash and stem grapes. Crush, add water, and bring to a boil. Simmer for 10 min. Extract juice and allow it to stand overnight in a cool place. Then strain through a double thickness of cheesecloth to remove tartrate crystals. Measure 4 cups into a kettle and add sugar. Follow standard procedure for making jelly without added pectin.

JELLY

With added pectin
Makes 9 8-oz. glasses

3 lb. grapes
1/2 cup water
7 cups sugar
1/2 bottle liquid pectin

Stem and crush grapes. Add water, bring to a boil, and simmer for 10 min. Extract juice. Measure 4 cups into a kettle and add sugar. Follow standard procedure for making jelly with liquid pectin.

FROZEN JELLY

Makes 6 8-oz. jars

2 lb. grapes
4 cups sugar
2 tbs. water
1/2 bottle liquid pectin

Stem, wash, and crush grapes. If they are not very juicy, heat them slightly. Put in a jelly bag and squeeze out juice. Combine 2 cups with the sugar. In a separate bowl, mix water and pectin, and add to fruit. Stir for 3 min. Pour into clean, cold jars and seal. Let stand for 24 hr. at room temperature. Then put in freezer.

JAM

Without added pectin
Makes 6 8-oz. glasses

4 lb. grapes
6 cups sugar
1/2 cup water

Wash and stem grapes and measure out 8 cups. Squeeze out pulp. Simmer skins in water for 20 min. In a separate kettle, simmer pulp without water for 5 min., then press through a sieve. Combine deseeded pulp, skins, and sugar and bring to a boil. Follow standard procedure for making jam without added pectin.

JAM

With added pectin
Makes 11 8-oz. glasses

3 1/2 lb. grapes
7 1/2 cups sugar
1/2 cup water
1/2 bottle liquid pectin

Wash and stem grapes, slip skins and chop them. Add water to pulp and simmer for 5 min. Press through a sieve. Combine deseeded pulp with skins. Measure 5 cups fruit into kettle and add sugar. Follow standard procedure for making jam with liquid pectin.

BUTTER

Makes 11 8-oz. glasses

5 1/2 lb. grapes
1/2 cup water
7 1/2 cups sugar
1/2 bottle liquid pectin

Wash, stem and crush grapes. Bring to a boil with the water and simmer for 10 min. Extract juice and use it to make jelly. Press pulp through a sieve or food mill and measure 5 cups into a kettle. Mix in sugar. Then follow standard procedure for making jam with liquid pectin.

CONSERVE

Makes 8 8-oz. glasses

4 lb. grapes
1 orange
1 cup nuts
4 cups sugar
1 cup seedless raisins
1/2 tsp. salt

Wash and stem grapes and slip skins. Measure 4 1/2 cups pulp into a kettle and boil for 10 min. Remove seeds by pressing through a sieve. Chop orange fine without peeling. Chop nuts fine. Add orange, sugar, raisins, and salt to sieved grape pulp and boil hard, stirring constantly, till mixture starts to thicken—about 10 min. Add grape skins and boil, stirring constantly, until a jam test is passed. Immediately remove from heat and stir in nuts. Skim. Ladle into hot, sterilized glasses and seal.

PICKLED GRAPES

Makes 4 pints

4 lb. grapes
1 cup sugar
2 cups water
1 cup cider vinegar
1 tbs. chopped ginger root
1 1/2 tsp. whole cloves
2 pieces stick cinnamon

Use seedless grapes. Or buy about 5 lb. of slightly underripe purple or red, seedy grapes and cut out the seeds. Wash and stem grapes. Heat all other ingredients together till sugar dissolves. Then add grapes and simmer till tender. Pack in hot, sterilized jars and seal.

WINE VINEGAR

Follow directions for making vinegar out of apples. Instead of adding sugar to the grape juice after it has fermented for the first time, siphon it into the second fermenter and leave the top open until vinegar has been formed. The vinegar should then be pasteurized and bottled.

GREENS

Greens are leafy plants which are generally boiled or steamed and eaten as vegetables. They include beet greens, collards, dandelions, Good King Henry, kale, Malabar spinach, mustard greens, New Zealand spinach, spinach, Swiss chard, tampala, turnip greens, and watercress. All are preserved in the same way.

FREEZING

Use young, bright green leaves. Remove tough stems. Tear up very large leaves. Follow standard freezing procedure. Blanch approximately 1 lb. vegetable in 2 gal. water for 2 min. Blanch collards for 3 min. Drain thoroughly before packing in rigid containers. Spinach is often chopped fine before packaging. To serve, cook as follows:

Beet greens	6–12 min.
Collards	12–15
Dandelions	4–6
Good King Henry	8–15
Kale	8–12
Malabar spinach	8–15
Mustard greens	8–15
New Zealand spinach	8–15
Spinach	4–6
Swiss chard	8–10
Tampala	8–15
Turnip greens	15–20
Watercress	4–6

CANNING

Use young leaves; cut out tough stems. Place leaves in a kettle with very little water, cover, and steam until they are wilted. Turn leaves over frequently to prevent sticking, and cut through them with a knife several times. Pack loosely into jars to 1/2 in. of top. Add 1/4 tsp. salt to pints; 1/2 tsp. to quarts. Cover with boiling water to 1/2 in. of top. Process in pressure canner at 10 lb. pressure. Pints for 70 min.; quarts for 90 min.

GROUPERS

Groupers are warm-water fish found in the Atlantic and Gulf of Mexico. Wrap and freeze them in the usual way. Store for 6–9 mo.

GROUSE

Handle and freeze like pheasant.

GRUMICHAMA

The grumichama is a small, tropical evergreen tree which produces sweet, cherry-size fruits in summer. These are usually red but may be purplish-black or black.

JELLY

Makes 3 8-oz. glasses

5 lb. grumichama
4 cups sugar

Wash fruit and remove stem and blossom ends. Put in kettle and mash; then barely cover with water and simmer till soft—about 20 min. Extract juice and measure 4 cups into a kettle. Add sugar. Follow standard procedure for making jelly without added pectin.

GRUNIONS

Grunions are small, smeltlike fish which are popular in California from Monterey south. They have the happy habit, when night tides are very high, of literally coming ashore to lay their eggs. You can catch just about all you want in your hands.

Most people eat them then and there, but there is no reason why you shouldn't take them home and freeze them in aluminum foil or polyethylene bags. Store for 5–7 mo.

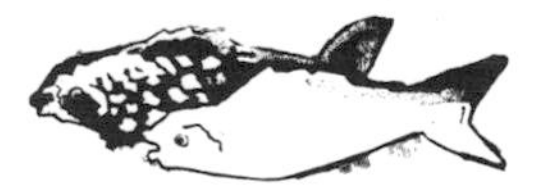

GRUNT

The white grunt and yellow grunt are noisy, smallish, salt-water fish found from the Carolinas south. Wrap and freeze them according to standard procedure. Store for 6–9 mo.

GUAVA

The guava is a tropical evergreen tree producing fruit which is usually too sour to eat fresh but which makes a superb cooked product. The fruits vary a great deal in appearance but generally have a yellow skin and white, yellow, pink, or red flesh. Fruits to be preserved should be picked when they are yellow but still firm.

FREEZING

Halves. Wash fruit, remove stem and blossom ends, and peel. Cut in half and scoop out seeds. Leave in halves or slice. Mix 4 cups fruit with 1 cup sugar. Or cover fruit with a sirup made of 1 cup sugar and 4 cups water. Package and freeze. Store for 10–12 mo.

Sauce. Cut washed, peeled, and seeded fruit into 1/4-in. slices and cook till just tender in a sirup made of 1 1/4 cups sugar to each 2 cups water. Remove from range, cool, crush, and ladle into cartons. Cover with the cooking sirup. Seal and freeze. Store for 10–12 mo.

CANNING

Halves. Wash, remove stem and blossom ends, and peel. Cut in half and scoop out seeds. Make a 40 percent sugar sirup and cook halves in this mixture for 2 min. Then pack in jars, adding sirup to each layer, to 1/2 in. of top. Cover with sirup. Leave 1/2 in. head space. Seal. Process in boiling water. Pints for 16 min.; quarts for 20 min.

Sauce. Wash, trim, and cut up fruit. Put through food mill. Simmer pulp till thick. Add 2 cups sugar to each 4 cups pulp if fruit is sour. Use somewhat less sugar for sweeter fruit. Cook rapidly for 10 min., stirring constantly. Pour into clean, hot jars to 1/2 in. of top. Seal. Process in boiling water. Pints for 5 min.; quarts for 8 min.

JUICE

Wash, trim, and slice ripe fruit. Put in a kettle and add 2 cups water to each 4 lb. fruit. Boil for 15 to 20 min. Drip through a jelly bag and strain through cheesecloth. You can then cool and freeze juice. Or you can reheat it to boiling, pour into Mason jars to 1/2 in. of top. Seal and process in boiling water. Pints for 5 min.; quarts for 8 min.

For a sweeter juice, add 1/4 to 1/2 cup sugar per quart of juice before reheating and processing.

JELLY

Without added pectin
Makes 4 8-oz. glasses

5 lb. guavas
4 cups sugar

Wash fruit, remove stem and blossom ends, and slice into a kettle. Add water barely to cover. Boil till fruit is soft—about 15 to 20 min. Crush, put in jelly bag, and extract juice without squeezing. Measure 4 cups into a kettle and add sugar. Follow standard procedure for making jelly without added pectin.

JELLY

With added pectin
Makes 9 8-oz. glasses

2 lb. guavas
2 1/4 cups water
1/2 cup lemon juice
7 cups sugar
1/2 bottle liquid pectin

Wash, trim, and slice fruits. Combine with water and cook till soft. Crush. Put in jelly bag and extract juice without squeezing. Measure out 3 1/2 cups and combine with lemon juice and sugar. Follow standard procedure for making jelly with liquid pectin.

BUTTER

Makes 5 8-oz. glasses

5 lb. guavas
6 cups sugar
6 tbs. lemon juice
3 tbs. grated fresh ginger root
3/4 tsp. ground allspice
3/4 tsp. ground cinnamon

Wash, trim, and slice fruits. Place in kettle and add water barely to cover. Boil till soft. Put through a food mill. Measure out

8 cups and combine with other ingredients. Cook slowly, stirring often, till thick. Ladle into hot, sterilized glasses and seal.

SHERBET

2 1/2 cups unsweetened guava juice
1 1/8 cups sugar
2 tbs. lemon juice
1 1/4 cups thin cream
2 egg whites
few grains salt

Use fresh, frozen, or canned guava juice. Combine 1 cup with 1 cup sugar and boil for 3 min. Chill. Add remaining guava juice, sugar, and lemon juice and freeze the mixture in a pan. Then spoon it into a chilled bowl and beat until fluffy. Add cream. Beat egg whites stiff, add salt, and add to mixture. Then pour mixture into pan and freeze firm.

GUINEA FOWL

The guinea fowl is a noisy, white-spotted, gray bird with a body about the size of a chicken. It is raised mainly in the South. It can be preserved like chicken.

HADDOCK

Haddock, once plentiful, has become so scarce that you're likely to have trouble either catching or buying it. If you come by any quantity of the fish, however, by all means wrap what you can't use right away in aluminum foil, put it in your freezer, and store for 7–10 mo.

Haddock is also smoked to make finnan haddie. Real finnan haddie is rarely encountered in the United States, however, and the Europeans are the only ones who seem to know how to make it. But you can produce a very satisfactory substitute by curing and smoking haddock according to the directions in Chapter 4.

HALIBUT

Halibut is a large, flat, salt-water fish which we especially like to bake in a rich, sherry-flavored egg sauce.

FREEZING

Fillet the fish or package them whole. Wrap in aluminum foil and freeze. Store for 6–9 mo.

CANNING

Clean and scale fish, cut off fins, tail and head, and wash thoroughly. Cut into pint-jar-length pieces without removing the backbone. Soak for 1 hr. in 2 qt. water and a generous 1/3 cup salt. Drain thoroughly. Pack solidly, but without crushing, into jars to 1/2 in. of top. Seal. Process in pressure canner at 10 lb. pressure for 110 min.

SMOKING

See Chapter 4.

SALTING

See Chapter 5.

RACKLING

Rackling is a Scandanavian fish product. It is easily prepared by washing and cleaning a halibut, removing the head above the collarbone, splitting the fish down the back and removing the backbone. Then cut the sides lengthwise into 1-in. strips which are held together at the collarbone. Wash thoroughly, taking care to get rid of all blood, and soak for 1 hr. in a brine made in the proportions of 3/4 cup salt to 1 qt. water. Then hang the fish outdoors in a dry, shady, breezy place to dry for about 2 wk. The fish is then stored in a cool, dark, dry place.

Rackling is eaten like venison jerky or may be freshened in water for several hours, steamed, and made into fish cakes.

HERRING

Like all fish, these small denizens of the Atlantic and Pacific oceans can be frozen, but the little 3- and 4-inchers are usually canned like sardines, while larger fish are cured and given a cold or hot (less good) smoke. Smoked fish that are left whole are called bloaters; those that are split are called kippers.

Herring can also be salted.

PICKLED HERRING

Clean, wash, and behead fish and place them in a crock. Cover with a mixture of 2 1/2 lb. salt, 2 qt. white vinegar and 2 qt. water. Store the crock in a refrigerator for 5 days; then pour off the pickle and soak the fish in cold fresh water in the refrigerator for 8 hr. Dry and fillet the fish and cut the fillets crosswise into 1/2-in. pieces.

In the meantime, make a pickle in the proportions of 3 cups water, 2 cups white vinegar, 1 cup sugar, 6 tsp. allspice, and 6 sliced medium onions. Combine these in a kettle and bring to a boil; then cool completely.

Pack fish pieces and onion slices in alternate layers in clean, sterilized jars to 1/2 in. of top. Cover with the pickle and put a piece of crushed plastic film in the tops of the jars to hold the fish under the liquid. Seal and store in the refrigerator. Do not use for at least 5 days. Store for up to 6 mo.

HICKORYNUTS

Hickorynuts are produced by the shagbark and shellbark hickory trees; and there is a new kind of hickorynut, called a hican, which is produced by a tree that is a cross between a hickory and pecan. In all cases, the nuts are mature when they fall free of the husks. They are then processed in the manner described in Chapter 8.

HIGHBUSH CRANBERRY

The highbush cranberry is a species of viburnum which is grown mainly for ornament. The shrub is also known as the American cranberrybush and Pembina. The clusters of red fruits make an excellent jelly, although they have an unpleasant odor, which is occasionally retained in the jelly.

JELLY

Makes 5 8-oz. glasses

2 lb. cranberries
6 cups water
3 cups sugar

If you use berries that are just beginning to turn color, the jelly will be a light yellow. Pink berries make pink jelly. Red berries make red jelly, but this may have too strong a flavor to be palatable.

Wash and stem fruit, crush, add water, and boil for 5 min. Extract juice. Measure out 4 cups, combine with sugar, and boil slowly to just below jelly stage (about 216°). Ladle into hot, sterilized glasses and seal.

BUTTER

Makes 5 8-oz. glasses

Measure pulp left from making jelly and combine it with 1/3 cup water for each cup pulp. Bring to a boil and boil for 4 min., stirring frequently. Then press through a food mill. Measure out pulp and combine it with an equal measure of sugar. Boil for 5 min. or a bit longer if you like a thicker butter. Pour into hot, sterilized glasses and seal.

HOGFISH

The hogfish is found mainly in the waters off Florida. It's an attractive red fish up to 2 ft. long. You can freeze it for 6–9 mo. It is also sometimes salted.

HONEY

As long as honey in the comb is protected from the air by beeswax, it can be stored as is at a temperature of 75° to 80°. But as soon as the honey begins to leak out, you should put the comb in a glass or plastic container and keep it covered. Store it at the same temperature.

To extract honey from the comb for storage usually calls for a special extractor. But you can do the job well enough by putting a piece of 3/4- or 1-in. galvanized mesh over a wire strainer. Place the honeycomb on the mesh and work it through with a wooden spoon, comb and all. Then let it drip through the strainer into a bowl with as little help as possible, though you will probably have to give it some encouragement to capture all the honey.

Let the honey stand in the bowl until bits of wax rise to the surface. Skim these off, add to the wax in the strainer, and heat in a low oven until the wax melts. Then let the wax cool and solidify. You can then drain off the remaining honey and add it to the strained honey.

Put the honey in the top of a double boiler and heat it to 155° for 30 min. to prevent fermentation and granulation. Then strain it into clean, sterilized Mason jars. Seal at once and cool. Store at 70° or below.

Use the beeswax to lubricate pulley stiles in double-hung windows, fill nail holes in wood paneling, and make candles and wax thread.

HORSE RADISH

The horse radish is a perennial plant which has long, thick, white roots that are used to make a fiery relish. The most tender, flavorful roots come from plants only one year old. The roots can be left in the ground the year round and dug up when you want them. Or you can store them in a root cellar at just above freezing.

To put up horse radish relish, cut the brown outer skin from the roots and grate the roots on a grater or cut them into cubes and chop them up in a blender. For each cup of grated root, add 1/2 cup white vinegar and 1/4 tsp. salt. Blend well. Pack in sterilized jars or bottles, seal tightly, and store in the refrigerator.

HUCKLEBERRIES

Huckleberries are seedier than blueberries and most people rate them inferior. But this is a little unfair. If you live in an area where huckleberries abound, by all means beat the birds to them and have yourself a feast. The berries can be preserved like blueberries.

JABOTICABA

The jaboticaba is a tropical evergreen tree bearing purple or black fruits more or less throughout the year. These are about the same size and shape as muscadine grapes and they taste like a delicious, juicy Concord grape.

Preserve jaboticabas by freezing or making them into jelly or jam. Follow directions given under *Grapes.*

JAMBOLAN

In Hawaii, the jambolan is called the Java plum. The tree is a massive, tropical evergreen with juicy, purple fruits like little, somewhat misshapen plums. They are often too sour to eat fresh, but they make good jelly.

JELLY

Makes 7 8-oz. glasses

2 qt. jambolans
2 1/2 cups water
1/2 cup lemon juice
7 cups sugar
1/2 bottle liquid pectin

Wash and stem fruit and put in a kettle with 1 1/2 cups water. Cook till fruit is soft —about 20 min. Extract juice. Let sediment in juice settle before measuring out 2 cups clear liquid. Put this in kettle with 1 cup water, lemon juice, and sugar. Follow standard procedure for making jelly with liquid pectin.

JERUSALEM ARTICHOKE

The Jerusalem artichoke is the warty tuber of a tall perennial sunflower. It is usually boiled or steamed and served as a vegetable. In warm climates, where the soil does not freeze, leave the tubers in the ground and dig them up throughout the winter when you need them. In cold climates harvest the tubers in the fall and store them in sand at 33° to 40°. The alternative is to dig a 2-ft.-deep trench, place the tubers in this, and cover them with an inch or two of soil or sand. Then fill the trench to the top with leaves, salt hay, etc.

JEWFISH

Jewfish are huge, sluggish, salt-water members of the bass family. They are gaining in popularity as food fishes but it will be a long time before they are ranked ahead of a lot of smaller fishes. Even so, if you cut out steaks, you can freeze them in aluminum foil, and you'll undoubtedly enjoy them. Store for 6–9 mo.

JUJUBE

The jujube is a deciduous tree producing small reddish-brown fruits which closely resemble dates. Since the tree is widely grown in China, it is also called the Chinese date. In the United States it is grown mainly in Southern California.

DRYING

Just leave the fruit on the tree until the skin wrinkles and turns a dark brown. Leave it on the tree for a few more days to dry further in the sun. Then pick it and store it in reasonably tight containers in a dark, dry, cool place.

CANDIED JUJUBES

See Chapter 10. Use dried fruit or fresh fruit which is reddish-brown in color. Puncture the skin all over with a sharp fork, knife—anything that is handy—before cooking in sirup.

JUNEBERRIES

Juneberries are close relatives of saskatoons; and the sweet, juicy, purplish-black berries are preserved in the same way.

KAFFIR

Named after the African tribe that first grew it, kaffir is a cereal grain now fairly widely grown in the United States. The large seed head is suggestive of an ear of husked popcorn, but unlike popcorn, each seed is wrapped in a hull and there is no cob.

After threshing and winnowing, the grains can be ground into flour and used for making bread.

KALE

Kale is an annual plant that grows almost everywhere. The crinkled leaves are cooked as greens. For how to preserve them, see *Greens*.

KINGFISH

The northern kingfish is also known as whiting, which confuses it to a certain extent with the silver hake, to which the whiting name is more commonly applied. On the Pacific Coast the regional kingfish variety is known as corbina.

Caught in salt water, kingfish have few bones and a fine flavor. Freeze them in aluminum foil and store for 6–9 mo. Corbina is often made into escabeche (see *Mackerel*).

KIWIS

Also called Chinese gooseberries, kiwis are ugly vine fruits grown mostly in New Zealand but to some extent in Southern California. They are about the size and shape of a hen's egg and are covered with fuzzy brown skin. But the flesh is a pretty translucent green and has a delicious, unusual flavor and a high Vitamin C content.

FREEZING

Peel and cut the fruits into quarters or thick slices. Pack in plastic containers and cover with 40 percent sugar sirup. Store for 10–12 mo.

CANNING

Peel kiwis and cut into quarters or leave whole. Pack in jars without crushing and cover with 30 or 40 percent sugar sirup. Leave 1/2 in. head space. Seal. Process in boiling water. Pints for 20 min.; quarts for 25 min.

JAM

Makes 3 8-oz. glasses

2 lb. kiwis
1 cup water
1 lemon
3 cups sugar

Cut fruits in half, scoop out pulp and combine in a kettle with the water and juice of the lemon. Bring to a boil and simmer for 10 min. Crush. Add sugar and boil, stirring frequently, to jam stage. Ladle into hot, sterilized jars and seal.

KOHLRABI

Kohlrabi produces a small bulb which looks and tastes very much like a white turnip but which appears just above the ground, rather than below. To freeze, cut off tops and roots. Peel and dice the bulbs in 1/2-in. cubes. Follow standard freezing procedure. Blanch for 1 min. Store for 10–12 mo. To serve, boil for 8–10 min.

KUMQUATS

PRESERVES

Makes 3 8-oz. glasses

1 lb. kumquats
1 cup water
2 cups sugar

Wash kumquats and cut a small gash across each one. Cover with water, bring to a boil, and simmer for 5 min. Drain. Combine sugar and 1 cup water and bring to a boil. Add kumquats and boil for 10 min. Cover and let stand overnight. In the morning, boil uncovered for 10 min. more. Let stand overnight again. Then bring to a boil once more and cook until the fruit is clear and the sirup is thick. Pack into hot, sterilized jars. Cover with sirup. Seal.

JELLY

Makes 2 8-oz. glasses

1 lb. kumquats
1 lb. sugar
3 cups water

Wash and slice kumquats and boil for 15 min. in water. Cover and let stand overnight. In the morning boil for 5 min. and let stand for 1 hr. Extract juice. Combine with sugar and boil to jelly stage. Ladle into hot, sterilized glasses and seal.

MARMALADE

Makes 8 8-oz. glasses

1 3/4 lb. kumquats
2 1/4 cups water
1/8 tsp. baking soda
5 cups sugar
1/2 bottle liquid pectin

Cut kumquats in half and remove seeds; then chop coarsely. Combine with water and soda, bring to a boil, and simmer for 15 min., stirring now and then. Measure 4 1/2 cups into a kettle and combine with sugar. Follow standard procedure for making jam with liquid pectin.

CANDIED KUMQUATS

Kumquats can be candied like grapefruit. The fruits are left whole, however. Gash each slightly. Boil only once to remove bitterness.

LAMB

When we moved to Lyme, Connecticut, Mrs. Susi, our realtor, told us how she and our other new neighbors made a practice of buying five or six lambs and grazing them in their fields. They then had them sheared and sent the wool to be woven into "the most beautiful blankets you've ever seen." And finally they had the lambs slaughtered, cut up, and frozen.

We got so excited that we were on the verge of going into the sheep-raising business ourselves. Then our youngest daughter Cary put her foot down and said no. And that was that. But when Cary and husband Charlie get transferred to Houston—we'll see.

FREEZING

Trim excess fat from the meat. Separate chops and other small cuts to be packed together with foil or wax paper. Wrap tightly in aluminum foil and place in freezer. Store for 8–10 mo.

Lamb kidneys are packaged in aluminum foil or rigid plastic containers and frozen for 4–5 mo.

SMOKING

Cure in salt or brine—thin cuts for 10 to 14 days; legs and shoulders for 25 to 30 days. Brine-cured meat must be smoked if it is to be stored for any length of time. Dry-cured meat need not be smoked, though it usually is.

Smoke lamb for about 48 hr. at 100° to 120°. Cool and wrap securely to protect against insects. Store in a cool, dry, well-ventilated place. Do not store much longer than 4 wk., however, because the meat dries out rapidly and develops a strong flavor. This is particularly true of thin cuts.

CANNING

Follow standard procedure for canning meat in a pressure canner. Cut-up meat which is preserved by either the hot-pack or raw-pack method is processed at 10 lb. pressure. Pints for 75 min.; quarts for 90 min.

To put up ground meat, remove excess fat and mix the meat with 1 tsp. salt per pound. Shape into thin patties and cook them in a 325° oven till medium done. Drain off fat and dry patties on paper towels to remove any fat that adheres to them. Then pack patties in jars to 1 in. of top and cover with boiling lamb broth to 1 in. of top. Process at 10 lb. pressure. Pints for 75 min.; quarts for 90 min.

LARD

Lamb or mutton fat may be rendered into lard and mixed in small quantities with pork fat. See *Pork.*

LEEKS

This mild-flavored member of the onion family is grown for its long, cylindrical stalks. The stalks can be harvested any time after they are 3/4 in. across, but those over 1 1/2 in. are the best.

DRY STORAGE

In mild climates, leave the leeks in the ground and harvest them as you need them. Elsewhere you can dig them up just before the ground freezes, stand them upright in a box, and pack soil around the roots and white portions of the stalks. Store in root cellar at 40° or below.

FREEZING

Freezing leeks is easy and preserves the stalks more reliably than dry storage. Cut off roots and trim the stalks to 8- to 12-in. lengths. Wash thoroughly. You may have to peel off several layers of flesh to remove all the little pockets of dirt. Follow standard freezing procedure. Blanch for 4 min. Package in small bundles in aluminum foil. Store for 10–12 mo. To serve, cook for 8–10 min.

LEMON BALM

This is also known simply as balm. It is a perennial herb with lemon-scented leaves. Collect and dry these before flowers appear, and store them in tight containers in a dark, dry place. You can use them to make a tea.

LEMONS

Note initial comment under *Oranges.*

JUICE

Extract juice with a reamer, not a press. Strain through cheesecloth to remove seeds and pulp. Pour into glass freezer jars (plastic might affect the flavor) and freeze. Store for 3–4 mo.

JELLY

Makes 5 8-oz. glasses

4 medium lemons
1 1/2 cups water
4 1/4 cups sugar
1/2 bottle liquid pectin

Grate rinds and ream juice from the pulp. Mix 1 tbs. rind, 1/2 cup juice, and water and let stand for 10 min. Strain through several layers of cheesecloth and measure 2 cups into a kettle. Add sugar. Follow standard procedure for making jelly with liquid pectin.

SHERBET

5 lemons
2 oranges
1 1/2 cups sugar
1 qt. water
1 beaten egg white

Squeeze juice from fruit and strain out seeds. Boil sugar in 2 cups water until it dissolves; cool. Add remaining water and juice; mix well. Pour into a pan and freeze to mush in the refrigerator. Then beat well in an electric mixer and fold in egg white. Return to refrigerator, and freeze firm.

PECTIN

Use peels of lemons from which juice has been extracted. Grate off the yellow skin and scrape off pulp clinging to the white peel. Cut peels into small strips and place 4 packed cups in a kettle with 2 qt. water and 1 tbs. tartaric acid (from the drugstore). Let stand for 2 hr. Measure depth of mixture and boil until depth is reduced 50 percent. Strain into a bowl through four or five thicknesses of cheesecloth. Then place rind in 2 qt. boiling water and 1 tbs. tartaric acid; boil down by half once more; and strain. Then repeat process a third time.

Mix the three extractions together, bring to a boil, and pour into hot, clean Mason jars to 1/2 in. of top. Seal. Process in simmering water to cover for 30 min.

This pectin can be used instead of commercial pectin at the rate of 1/4 cup to each cup of fruit juice.

LEMON VERBENA

Lemon verbena is a deciduous shrub which can be grown outdoors the year round only in our warmest climates. Elsewhere it is grown in greenhouses. The extremely fragrant leaves are picked and dried to make a tea. Store in tight containers in a dark place. The leaves must be allowed to steep for 10 min. or more to give up their full flavor.

LENTILS

Lentils are annual plants producing flat pods which contain small, lens-shaped, brown seeds. They make a delicious soup and are also sometimes baked. To preserve them, let the pods and seeds dry on the plants. Then remove the seeds and store them in tight containers in a cool, dark, dry place.

LIMES

Note initial comment under *Oranges.* However, limes are best stored at 45° to 50°.

JUICE

Extract juice by reaming. Strain through cheesecloth to remove seeds and pulp. Pour into glass freezer jars and freeze. Store for 3–4 mo.

JELLY

Makes 8 8-oz. glasses

8–10 medium limes
2 1/2 cups water
5 cups sugar
1 bottle liquid pectin
green food coloring

Squeeze out and strain juice. Measure 1 cup into a kettle; add water, sugar, and a few drops of food coloring. Follow standard procedure for making jelly with liquid pectin.

SHERBET

6 medium limes
1 lemon
1 qt. water
1 1/2 cups sugar
3 beaten egg whites

Squeeze limes and measure 1 cup juice. Squeeze lemon and combine with lime juice. Heat water and dissolve sugar in it; then cool. Combine with fruit juice. Add more sugar if you like a sweeter sherbet. Pour into pan and freeze to a mush in the refrigerator. Then add egg whites and mix well. Freeze firm.

LINGCOD

The lingcod is a codlike, salt-water fish averaging about 5 lb. Freeze it in aluminum foil and store for 6–9 mo. It can also be smoked or salted.

LOBSTER

FREEZING

Drop lobsters into boiling salted water and boil for 20 min. Chill in cold salted water until you can handle. Then pick out meat and chill it as rapidly as possible in the refrigerator or a bowl placed in cold water. Pack meat in rigid plastic containers. Freeze. Store for 3 mo. Ready to serve when thawed.

CANNING

Cook, chill, and pick out meat as above. Drain meat, then dip it briefly in 1 gal. water to which 1/2 cup vinegar or lemon juice has been added. Press out moisture. Pack 3/4 cup meat tightly, but without crushing, into each half-pint Mason jar. Cover with a hot brine made of 1 gal. water and 3 tbs. salt. Leave 1/4 in. head space. Seal. Process in pressure canner at 10 lb. pressure for 70 min.

LOGANBERRIES

Loganberries are closely related to the trailing forms of blackberry. They are grown mainly on the Pacific Coast. The big red berries are put up in the same ways as blackberries (which see). The only slight difference is that, in making loganberry jam with liquid pectin, you should use only 6 1/2 cups sugar.

LOQUATS

The loquat is a tropical evergreen tree bearing clusters of oval, yellow, or orange fruits about the size of large olives. The flesh is juicy and either sweet or a little acid.

CANNING

Whole. Wash fruit, remove stem and blossom ends and peel. Cut out seeds or not, as you prefer. Cook 3 min. in 40 percent sugar sirup. Pack in jars to 1/2 in. of top and cover with sirup. Leave 1/2 in. head space. Seal. Process in boiling water. Pints for 15 min.; quarts for 20 min.

Sauce. Wash fruit, remove stem and blossom ends and seeds. Then chop. Mix 1 cup water with 4 cups pulp and cook till tender. Mix in 1 1/2 cups sugar for each 4 cups pulp and cook for 5 min. more. Pack in Mason jars to 1/2 in. of top. Seal. Process in boiling water. Pints for 15 min.; quarts for 20 min.

JELLY

Use underripe fruit. Wash, remove stem and blossom ends and seeds. Just cover with water, bring to a boil, and simmer till tender. Extract juice. Combine 3 cups with 2 cups sugar in a kettle. Add 2 tsp. lemon juice if fruit is sweet. Boil to jelly stage. Ladle into hot, sterilized glasses and seal.

LOVAGE

A tall perennial that grows in temperate climates, lovage has celery-flavored leaves which are used to season soups and salads. Use young leaves only. These can be frozen without blanching in small, airtight polyethylene bags. They can also be dried and stored in bottles.

LYCHEE

Also called leechee and litchi, this is a tropical evergreen tree that produces large clusters of 1- to 1 1/2-in. fruits with rough red or yellow skins and succulent white flesh. Freeze the fruits as soon as they are ripe, and don't waste time getting them from tree to freezer. Just wash, put in polyethylene bags or rigid containers, seal and freeze. Store for 10–12 mo.

You can also remove shells; pull the halves of the fruit from the seed; and pack in 50 percent sugar sirup before freezing.

MACADAMIA NUTS

The macadamia is a tall, tropical evergreen tree bearing sweet round nuts throughout much of the year. When the nuts are mature, they fall to the ground. Harvest them fairly frequently to prevent spoilage, and remove the husks promptly. Then spread the nuts out in wire trays in a shady, airy place to dry for 2 to 3 wk. You can then store the nuts in sacks in a dry, airy, cool place for at least 4 mo. Handle shelled nuts by the methods described in Chapter 8.

MACKEREL

This handsome blue salt-water fish can be frozen in aluminum foil but should not be stored for much over 6 mo. because of its oily flesh. The fish are also smoked or salted.

ESCABECHE

This is a popular and very old Spanish method of preserving fish. Proportions given are for 10 lb. fish.

Fillet fish and cut into small serving portions. Wash well, drain, and immerse for 30 min. in a brine made of 3/4 cup salt per quart of water. Dry on paper towels. Heat 1 pt. olive oil in a frying pan and add the fish, a clove of garlic minced, 6 bay leaves and a good dash of red pepper. Cook until fish is light brown; then remove it from the pan to cool.

To the hot oil, add 1 cup sliced onions, 1 tbs. black peppercorns, 1/2 tbs. cumin seed, 1/2 tbs. marjoram, and 1 qt. white vinegar. Cook slowly for about 30 min. Then cool.

When the fish are cold, sprinkle with 1 tsp. red pepper and add a dozen bay leaves. Pack in sterilized Mason jars and cover to 1/4 in. of top with the cold oil mixture. Seal. Store in the refrigerator.

MAHIMAHI

Of all the pleasant experiences we have had collecting information for this book, none has been quite so pleasant as our mahimahi research. We knew mahimahi was a fish that we had enjoyed eating in Hawaii, but what kind of fish we couldn't remember for sure; and despite inquiries, we couldn't seem to find out. Then we recalled that the last time we had eaten mahimahi was at the Hilo Hotel in Hilo, and we dropped a line to the manager asking for help.

Her reply came within a fortnight. It appeared that one of the hotel's long-time waitresses, Mary Dominquez, had undertaken to research the matter for us; and the report she typed would fill the better part of a page in any encyclopedia.

Mahimahi, we learned, is the salt-water fish (*not* the mammal) more commonly known as dolphin. There are two species, one up to 70 inches long, the other less than half that size. They are swift, playful fellows that swim with an undulating motion and frequently leap entirely out of the sea (the islanders believe this means that bad weather is brewing). They can see a trolled lure from a distance of over 100 ft., fight like furies when hooked, and are delicious to eat—especially, we like to think, when Mary is serving.

Since the flesh is a bit dry, it is advisable to keep it from drying out further in the freezer by glazing the fillets before wrapping them in aluminum foil or polyethylene bags. Prepare like any other fish fillet. Store for 4–6 mo.

MALABAR SPINACH

Malabar spinach is a small annual vine with large leaves which are sometimes used as a substitute for spinach. For information on how to put them up, see *Greens.*

MANDARINS

Mandarins include tangerines and satsumas. For how to store fresh fruit, see the initial comment under *Oranges.* You can also freeze mandarins by peeling, sectioning, and removing the white veins. Then pack in rigid containers and cover with 30 percent sugar sirup. Seal and freeze. Store for 6 mo.

TANGERINE MARMALADE

Makes 3 8-oz. glasses

3 tangerines
1 lemon
2 1/4 cups sugar

Wash and quarter fruits and remove seeds. Chop. Measure into a kettle and add 4 cups water for each cup pulp. Bring to a boil, then simmer until volume is reduced 50 percent—about 45 min. Let stand overnight to extract pectin from the lemon. The next morning measure out 3 cups pulpy juice, combine with sugar, and boil to jelly stage. Let cool to 190°. Ladle into hot, sterilized glasses and seal.

MANGOES

DRY STORAGE

Mangoes will keep for about 20 days if stored in a dark, very humid place at 50°.

FREEZING

The best varieties for freezing include Wootten, Fairchild, Pirie, Hansen, Ono, Joe Welch, and Hotoke. The fruits should be fully ripe. Wash, peel, and cut off and discard a slice of the stem end. Cut off the cheeks of each fruit. Don't cut close to the seed or slice the fruit as you do a peach, because the flesh near the seed is often fibrous.

If packing in sugar, mix 1 lb. sugar with 8 to 10 lb. fruit and let stand for a few minutes till sugar dissolves. Then put in containers. If packing in sirup, put fruit in containers and cover with 30 or 40 percent sugar sirup. Store for 10–12 mo.

SAUCE

1 1/2 lb. mangoes
1 1/2 cups water
1/2 cup sugar

Use ripe or green mangoes. Wash and peel and cut into pieces. Combine with water and cook till tender—about 15 min. Press through a food mill to remove stringy portions. Stir in sugar. Cool thoroughly. Package in rigid freezer containers and freeze. Store for 10–12 mo.

CANNING

Wash and peel fruit and slice into hot 30 or 40 percent sugar sirup. Let stand for 2 min. Pack fruit in jars to 1/2 in. of top. Boil sirup for 5 min. or a little more. Strain over fruit to 1/2 in. of top. Seal. Process in boiling water. Pints for 15 min.; quarts for 20 min.

Green mangoes can be canned in the same way.

JAM

Without added pectin
Makes 4 8-oz. glasses

3 lb. mangoes
3 cups sugar
1/2 cup water

Use half-ripe or ripe fruit. Wash, peel, and slice. Measure 6 cups into a kettle and cook with water until tender—about 15 min. Press through a food mill. Add sugar and simmer. Follow standard procedure for making jam without added pectin.

JAM

With added pectin
Makes 10 8-oz. glasses

5 or 6 lb. mangoes
2 tbs. lemon juice
7 1/2 cups sugar
1 bottle liquid pectin

Use fully ripe fruit. Wash, peel, slice, and crush fruit. Measure 4 cups into a kettle. Add lemon juice and sugar. Follow standard procedure for making jam with liquid pectin.

SHERBET

3 lb. mangoes
1 1/2 cups sugar
3/4 cup water
1/3 cup lemon juice
3 cups milk

Wash, peel, and slice mangoes, and press through a sieve or purée in a blender. Bring sugar and water to a boil and then let it cool. Combine with 3 cups mango purée and the lemon juice. Add this mixture slowly to milk, stirring constantly. Pour into pans and freeze to a mush. Then beat in a bowl and return to pans. Freeze firm.

MAYHAWS

Mayhaws are the small red berries of a hawthorn tree called the mayhaw *(Crataegus aestivalis)*. Two other hawthorns, Arnold hawthorn *(C. arnoldiana)* and Downy hawthorn *(C. mollis)*, also have edible red fruits.

JELLY

Makes 10 8-oz. glasses

3 lb. mayhaws
4 cups water
1/4 cup lemon juice
7 1/2 cups sugar
1/2 bottle liquid pectin

Wash and crush fruit; add water; bring to a boil and simmer for 10 min. Extract juice. Measure out 4 cups. Add lemon juice and sugar. Follow standard procedure for making jelly with liquid pectin.

SAUCE

Use fruits that are just beginning to show color. Wash. Combine 1 cup mayhaws with 1 1/2 cups water and cook till soft. Press through sieve. Combine 1/2 cup sugar with each cup pulp. Cook to desired thickness. Ladle into Mason jars to 1/2 in. of top. Seal. Process in boiling water. Pints for 5 min.; quarts for 10 min.

JUICE

Prepare fruits as for sauce. When cooked soft, extract juice in jelly bag and strain through cheesecloth. Bring to a boil. Pour into hot, clean jars. Seal. Process in boiling water. Pints for 5 min.; quarts for 10 min.

MELONS

Most of the melons grown in the United States are popularly known as cantaloupes; but just to set the record straight, they are nothing of the sort. The true cantaloupe is rarely grown in this country. What is called a cantaloupe is actually a muskmelon—which somehow isn't a very attractive name for such a delicious fruit.

The other melons grown in the United States are slow-growing winter melons. These include the casaba, Crenshaw, honeydew, and Persian melon, and they are every bit as delicious as muskmelons and can be stored two or three times as long (up to about 6 wk.) if held in a dark, humid place at 40° to 50°.

FREEZING

Use firm, well-ripened fruits. Cut in half, remove seeds, and scoop out melon balls. Or pare the fruits and cut the flesh into 3/4-in. cubes. Pack into rigid containers to 1/2 in. of top and cover with 30 percent sugar sirup. Seal and freeze. Store for 10–12 mo.

Some people also pack melon balls in orange juice.

MUSKMELON PICKLES

Makes 2 pints

1 medium melon
1 qt. cider vinegar
2 cups water
1 tsp. mace
2 3-in. pieces stick cinnamon
2 tbs. ground cloves
4 cups sugar

The melon should be no more than medium ripe. Cut it in half, remove seeds, peel, and cut the flesh into 1/2-in. slices. Combine with vinegar, water, and spices (in a bag) and bring to a boil. Place fruit in a hot nonmetallic container, cover with the boiling liquid, and let it stand overnight. In the morning, drain liquid into a saucepan, add cloth bag of spices, and bring to a boil. Then add sugar and fruit and simmer until fruit becomes translucent—about 1 hr. Remove fruit from liquid and pack into hot, sterilized jars standing in hot water. Bring sirup to a boil once more and cook it down to medium thickness. Discard spices. Pour sirup over fruit and seal jars.

MILK

FREEZING

Pasteurized and homogenized milk can be frozen and stored for 2–3 mo. Provide 1 1/2- to 2-in. head space in the container to allow for expansion.

Pasteurized heavy cream can also be frozen and will keep for 3–4 mo. For slightly longer storage, add 1 part sugar by weight to 10 parts cream by weight. Allow about 8 to 12 hr. for the cream to thaw in the refrigerator. It may then be whipped.

Sweetened cream can be whipped before freezing and frozen in mounds on a cookie sheet or other flat surface. Then pack into rigid plastic containers with aluminum foil between layers.

BUTTER

Start with day-old cream that has been chilled to 55° to 60°. After washing the churn well and rinsing it with cold water, fill it about halfway—no more—with the cream and start churning. Work at a steady pace—not too fast; not too slow. After the first ten strokes, open the churn to release carbon dioxide. Do this again after the next 20 strokes. From then on, continue churning for about a half hour until particles of butter about the size of a pea have formed.

Drain off the liquid and fill the churn with the same amount of cold water. Turn the churn at high speed for a few minutes. Drain, refill with water and churn some more. Then collect the butter in a strainer, and squeeze it with your hands or a paddle to remove remaining moisture. Then wash it well with ice water and put it in a mold.

An alternative way to make butter is recommended by our friend Marian Taylor, who grew up in Alabama. It yields good old-fashioned buttermilk as well as butter. In this case, you start with raw whole milk and let it stand in a warm room for about 24 hr. until it clabbers. Then put it in a churn and churn until butter forms. Pour off the buttermilk, fill the churn with cold

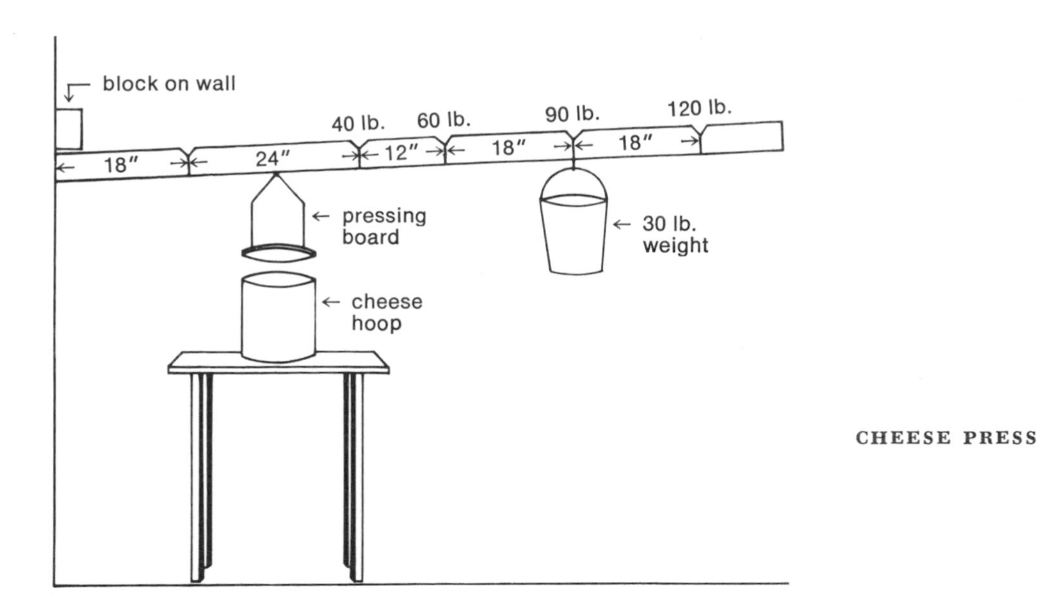

CHEESE PRESS

water and carry on as in the preceding instructions.

For salted butter, add salt before working the butter to remove moisture.

To pasteurize cream before making butter, heat water in the bottom of a stainless steel or aluminum double boiler to about 165°. Pour cream into the top of the double boiler, place over the hot water and heat the cream to 165°. Stir gently now and then during the process. As soon as the temperature has been reached, remove the cream from the heat and chill it as rapidly as possible to 50° by placing the pan in a large bowl filled with cold running water.

Store butter in the coldest part of the refrigerator but only for about 3 wk. For long-term storage, wrap butter in two layers of freezer wrap and store in the freezer for 6–9 mo. Only butter made with pasteurized cream should be frozen. Sweet butter does better than salted butter.

CHEDDAR CHEESE

Cheese is more difficult to cure than to make, and since proper curing conditions are not generally found in the home, cheese-making is not a very popular family pastime. But the following method for making a cheddar-type cheese was developed by the Department of Agriculture strictly for farmers and home owners.

The unusual equipment required is as follows: A floating dairy thermometer and rennet extract or rennet tablets—all available from a dairy-supply house. A 35-qt. canning kettle. A wire egg whip with crossed wires. A large piece of open-weave muslin. A large (about 1 gal.) tin can or pail with smooth straight sides and a dozen holes punched in the bottom with a 2-in. nail. This is called a cheese hoop. And a cheese press like that illustrated. The cheese press is made out of a 2 x 4 a little over 7 1/2 ft. long and a bucket filled with metal or rocks to weigh 30 lb. Actual pressing is done by a maple board. This is cut in a circle so that it just fits in the cheese hoop, and it is fastened to the bottom of the 2 x 4 by a short piece of 2 x 4.

Use whole cow's milk. Depending on the butterfat content, 50 lb. milk (almost 6 gal.) will produce from 4 to 6 lb. cheese. Half of the milk should come from the milking of the previous evening. This should be held overnight at 60°. The other half of the milk comes from the milking of the morning you start your cheese-making. (Using all evening milk or all morning milk results in inferior cheese.)

The first step in making cheese is to mix the milk in your cleaned canning kettle and heat it, stirring slowly but steadily. When it reaches 86°, remove it from the range and add enough rennet to coagulate the milk into a firm, jellylike curd in about 30 min. If using rennet extract, dilute 1 to 2 tsp. in a cup of cold water. If using rennet tablets, follow the maker's directions. Stir the rennet into the warm milk for 3 min., then cover the kettle, and let the milk stand undisturbed.

To determine when the curd reaches the proper jellylike consistency, place the back of your hand on it near the side of the kettle and press down. If the curd breaks away cleanly from the kettle, it is ready to be cut.

For cutting use a sharp knife with a blade long enough to reach the bottom of the kettle. Slice the curd one way into 3/8-in.-wide strips and then slice it crosswise in the same way. Then stir your egg whip through the curd for a couple of minutes or until the curd is cut into fairly uniform 3/8-in. cubes. Then continue stirring with a wooden spoon for about 15 min., when enough whey should have been expelled to separate the cubes of curd.

Return the kettle to the range and heat it very gradually until the curd reaches 100°. Try not to increase the temperature more than 2° every 5 min. Stir slowly and steadily to prevent the cubes from forming larger lumps.

When the curd reaches 100°, continue cooking at this temperature—stirring only occasionally—until the curd is firm. "At this time," says the USDA bulletin, "a handful of curd gently squeezed and suddenly released should easily break apart and show very little tendency to stick together. From 1 1/2 to 2 1/2 hr. after the addition of the rennet the curd should have this characteristic."

Remove the kettle from the range, let the curd settle for a few minutes, and dip off as much of the whey as possible. (This can be thrown out or fed to chickens.) Pour the rest of the whey and the curd into a large pan covered with cheesecloth. Let the curd drain thoroughly, and return it to the kettle. If any whey remains, drain it off.

Stir the curd until it cools to 90°. At this point it should have a rubbery texture and make a squeaky sound when chewed.

Add 3 3/4 tbs. dairy or kosher salt to the curd and mix it in evenly until it is dissolved and the curd is cooled to about 85°. It is then ready for hooping and pressing.

Put the hoop on a table under your press and place a circle of muslin in the bottom. Pour in the curd and cover it with another circle of muslin. Then pull down the press and adjust the weight to apply 40 lb. pressure for 10 min. Then increase to 90 lb. pressure for another 30 min.

Remove the cheese from the hoop, take off the cap cloths, and dip the cheese in 100° water to remove any fat adhering to the surface. Cut a strip of muslin 2 in. wider than the height of the cheese and about 1 in. longer than the circumference. Center the cheese on this and wrap it tightly. This is called bandaging the cheese. Replace the cap cloths, and slide the cheese back into the hoop. Press it for about 16 hr. under 120 lb. pressure.

At the end of the pressing period, remove the cloths and examine the cheese. If there are any openings in the surface, dip the cheese in 100° water again, replace the bandages, slide the cheese back into the hoop, and press it a while longer until the rind is sound. (Openings in the surface of a cheese allow molds to get inside.)

When you are satisfied with the condition of the surface, remove the bandages, wipe the cheese with a dry cloth, and put it for 3 to 6 days in a clean, dark, tightly screened place at 50° to 60°. If the humidity in the curing area is high, the cheese need not be covered in any way; but if the humidity is low, the bandages should be put back on the cheese to retard drying of the rind. During the curing process, turn the cheese over every day and wipe it off with a dry cloth.

When the surface of the cheese feels dry, dip one half and then the other half in melted paraffin so that it is completely coated. The paraffin should be hot enough (about 210°) so that it forms only a thin coating on the cheese. If it is not hot enough, it will form a thick coating, which scales off easily. If the paraffin doesn't stick to the cheese, it is because the surface is not dry and the cheese needs to be cured a little longer.

The final step in the cheese-making process is to return the cheese to the curing room to ripen for several months. If the room temperature is 60°, 4 mo. of curing is adequate if the cheese is of very high quality. But generally 6 mo. at 40° to 50° gives better results.

An alternate method of handling the cheese after it has been pressed sufficiently to form a sound surface is to wipe it off with a dry cloth and wrap it tightly in any plastic film that can be heat-sealed easily (cellophane, for example). Then place it in the curing room for 4 to 6 mo. Because the cheese is not dried before wrapping, it does not form a rind.

COTTAGE CHEESE

Cottage cheese is quite perishable even under refrigeration, but its storage life can be greatly improved by the following manufacturing method.

Raise or lower 1 gal. pasteurized skimmed milk to a temperature of 70°. Stir in 1/2 cup fresh commercial buttermilk and let stand at 70° for 16 to 24 hr., or until it clabbers. At this point, the curd should break away from the side of the kettle when you press it lightly with a spoon.

Cut the curd vertically into 1-in. squares with a knife; and then cut it into 1-in. cubes by running a small, stiff, U-shaped wire through it horizontally. Try to make the cubes as nearly uniform in size as possible.

After letting the curd stand for 10 min., add 2 qt. water warmed to 100°. Place the kettle on a rack in a larger kettle filled with water and heat the curd gradually to 100°. Hold at this temperature for 30 min. or a little longer. Stir gently with a wooden spoon every 5 min., being careful not to break up the curd.

When nearly all the curd has settled to the bottom of the kettle, remove it from the range. At this point, when you break the cubes between your fingers, they should break cleanly and look somewhat dry.

Pour the curd and whey into a large strainer and wash it with about the same amount of cold, fresh water as there was whey. This cools the curd and removes some of the acid taste. Let the curd drain until it is losing very little liquid, but don't let the surface dry out.

Make a brine by dissolving 2 1/2 oz. salt in 3 qt. water and pour a little into a large glass jar, rigid plastic container or even a sturdy plastic bag. Pour in the cheese and cover it completely with additional brine. Then seal the container tightly and put it in your refrigerator. The cheese will keep for up to 3 mo.

When you are ready to eat the cheese, pour off the brine and rinse the cheese several times in cold fresh water. Place it in a large bowl and work in salt to taste or at the rate of 1 tsp. per pound of cheese. Then work in a little sweet or sour cream to make the cheese creamy.

One gallon of skim milk makes about 1 1/2 lb. cottage cheese.

ICE CREAM

Latest intelligence has it that ice cream—that old, old American favorite—is today being consumed in this country in record quantities. Some 800 million gallons were sold by commercial producers in 1971 alone. Heaven only knows how many additional millions of gallons were made at home.

Yes, despite a dairy authority we consulted, Americans still make ice cream at home. He contended this is a nonsensical thing to do because commercial ice cream is so much better and so much cheaper than anything a family can crank out. He's probably right about the cost; and certainly no one can be foolish enough to pooh-pooh the quality of much of the ice cream sold in various and sundry emporiums. But there's something about home-made ice cream: the fun of licking the dasher.

The pleasure of that first spoonful pushed ever so slowly under the upper lip.

And then there's the practical side of the matter: Ice cream is a great way to preserve not only milk but also many fruits. If kept in tight plastic containers, it can be stored for a couple of months in the freezer, though it is best if eaten within about 3 wk.

Make ice cream in a hand-cranked or electrically driven ice-cream freezer. It is much better than ice cream made in a refrigerator freezer tray.

Make sure the freezer can is spotlessly clean and free of odors. Fill it no more than two-thirds full to allow for expansion. Drop in the dasher, put on the lid, and center the can in the tub under the gearcase.

Pack the tub as tightly as possible with 5 parts ice and 1 part rock salt. (These are not mixed before going into the tub. Just put in some ice; then some salt; and so on.) Add 1 cup cold water to dissolve the salt, and start cranking.

It takes about a quarter hour to turn the custard base into ice cream. If you are making vanilla ice cream, just keep cranking until the freezer turns hard. If you are making a fruit ice cream (the only other kind of ice cream we think a book on preserving should talk about), stop cranking after 7 or 8 min., stir the fruit carefully into the custard, and then continue cranking until the freezer turns hard.

The final step—if you can resist consuming the ice cream right then and there—is to pack it into rigid plastic containers to 1/2 in. of the top, seal, and place in the food freezer for storage.

Our basic custard for making ice cream comes from Elizabeth's Cousin Dolly. Do you mind if we say it is beautiful?

1 cup milk
1 egg
1/2 cup granulated sugar
few grains salt
1 pt. heavy cream
1 tsp. vanilla extract

This makes 1 qt. ice cream. Scald 3/4 cup milk in a double boiler. Beat egg yolk and save the white for something else. Add remaining cold milk, sugar, and salt to egg and combine the mixture with the hot milk. Cook, stirring, until mixture coats the spoon—about 3 min. Remove from heat, strain, and chill. Stir in cream and vanilla; pour into freezer can and go to work cranking.

To make fruit ice cream, chill the fruit, then crush and add just enough sugar to make it mildly sweet (it may not need any at all). Combine 1 cup with the above recipe when the custard is partially frozen. (If you increase the custard recipe, increase the amount of fruit in proportion.)

Some of the familiar fruits you can use include apricots, bananas, blueberries, sweet and sour cherries, peaches, pineapples, and strawberries.

MINT

Most people are content with whatever mint they happen to come by, and there is nothing wrong with that: Mint in any form is a delight. But if you plant mint in your garden, you should consider which variety you like especially—spicy peppermint, sweet spearmint, apple-flavored applemint, orange-flavored orangemint or pineapple-flavored pineapplemint.

The leaves of all mints are easily dried in an airy place and packed, crushed or whole, in tight containers. Use them in tea, salads, or pea soup or make mint sauce for spring lamb.

JELLY

Makes 5 8-oz. glasses

1 1/2 cups mint leaves
2 1/4 cups water
2 tbs. lemon juice
3 1/2 cups sugar
green food coloring
1/2 bottle liquid pectin

Wash and crush leaves in a saucepan. Cover with water and bring to a boil; then remove from range, cover, and let the leaves steep for 10 min. Strain out 1 3/4 cups liquid into another pan, add sugar, lemon juice, and a few drops food coloring. Follow standard procedure for making jelly with liquid pectin.

MOOSE

Moose meat is not highly recommended for eating but why not preserve it anyway? Handle like venison. In mild weather, it is advisable to skin your moose at once in order to let the carcass cool rapidly.

MULBERRIES

Mulberries are pretty insipid fruits, but if you like them you can easily freeze them. Just wash and pick over. Mix 5 cups fruit with 1 cup sugar. Package, seal, and freeze. Store for 10–12 mo.

MULLET

Mullet is a rather oily, salt-water fish of good flavor. It can be frozen, smoked, or salted. The roe may also be dried or smoked like shad roe.

MUSHROOMS

If you're not a mushroom expert, stick to the cultivated product until you have studied enough books and gone on enough field trips with a true expert to be able to identify the wild mushrooms. True, relatively few of these are poisonous; but unless you are an experienced collector, there are no easy ways, no tests to help you distinguish between the safe and unsafe.

Preservation of wild and cultivated mushrooms is the same, however.

FREEZING

Eliminate mushrooms that are spotted or decayed. Sort according to size. Wash in cold water. Trim off stem ends. Cut mushrooms more than 1 in. in diameter in halves or quarters.

If you steam mushrooms, dip them first for 5 min. in 1 pt. water mixed with 1 tsp. lemon juice. This helps to prevent darkening. Then steam whole mushrooms less than 1 in. across for 5 min.; buttons or quarters, for 3 1/2 min.; slices for 3 min. Chill in cold water, pack in rigid containers, and freeze.

Another method is to fry small quantities of mushrooms in margarine until almost done. Allow to air-cool or set the pan in cold water. Then pack and freeze.

DRYING

Pick over and wash. Remove stems from very large caps. Spread on paper or cloth to dry in the sun. When mushrooms seem very dry, place on baking sheets in a 225° for about 1 hr. to finish the process. Then pack in clean, tightly sealed jars and store in a dark, dry, cool place.

Lacking sun, you can do the whole drying job in the oven at about 160°.

CANNING

Trim off stem ends and discolored parts. Wash well. Cut mushrooms more than 1 in. across in halves or quarters. Place in a strainer in a covered kettle over boiling water and steam for 4 min. Pack into jars to 1/2 in. of top. Add 1/4 tsp. salt to half pints; 1/2 tsp. to pints. To improve the color of the mushrooms, also add 1/4 tsp. ascorbic acid to half pints; 1/2 tsp. to pints. Cover with boiling water to 1/2 in. of top. Seal. Process in pressure canner at 10 lb. pressure for 30 min.

PICKLED MUSHROOMS IN OIL

Makes 2 pints

1/2 cup lemon juice
1 qt. water
6 cups small mushrooms
2 cups white vinegar
1 tsp. salt
1/3 tsp. tarragon
3 bay leaves
1/2 tsp. basil
2 cloves garlic
1 1/2 cups olive oil

Combine lemon juice and water; add mushrooms; bring to a boil and simmer for 5 min. Drain. Add vinegar, cover, and let stand for 12 hr. Drain and pack mushrooms in jars. Mix salt and spices and add to jars in equal amounts. Add 1 garlic clove to each pint or 1/2 clove to each half pint. Cover with olive oil. Seal. Process in boiling water for 20 min.

PICKLED MUSHROOMS IN WINE VINEGAR

Makes 2 pints

1/2 cup lemon juice
1 qt. water
6 cups small mushrooms
2 tbs. mixed pickling spices
1 qt. wine vinegar

Combine lemon juice and water; add mushrooms; bring to a boil and simmer for 5 min. Drain and pack mushrooms in jars. Put spices in a bag, place in kettle with vinegar, and bring to a boil. Simmer for 1 min. Remove spices. Pour vinegar over mushrooms to within 1/4 in. of top. Seal. Process in boiling water for 20 min.

MUSKELUNGE

The muskie is a big, tough fighter much sought after in Minnesota and Wisconsin lakes. If you bring one to net, you can wrap and freeze it in the usual way. Store for 4–6 mo.

MUSSELS

Mussels abound in both fresh and salt water. They have long, thin, blue-black shells filled with delectable orange meats. Although they are not often preserved, they can be frozen, canned, or smoked like clams. They can also be pickled like oysters.

Salt-water mussels should be harvested only in cold weather. In warm weather they are poisonous. Many states establish quarantine periods on these mussels.

MUSTARD

Mustard is a member of the cabbage family grown for its leaves. For how to preserve, see *Greens.*

MUSTARD SEED

Dry and prepared mustards are made from seeds of the black mustard plant *(Brassica nigra)* or white mustard plant *(B. alba).* Both are 4-ft. annuals with clusters of yellow flowers followed by 1-in. seed pods. To collect the seed, cut off the seed stalks when the lowest pods start to dry and let the pods ripen on cheesecloth in a warm, dry place. Then shake or thresh out the seeds.

The whole seeds can be put in bottles and used to flavor potatoes, salad, stews, or coleslaw. They are also used in pickles. The easiest way to grind the seed at home is in a meat grinder with the fine blade or with mortar and pestle.

MUTTON

Handle mutton like lamb.

In Norway, leg of mutton is often brine-cured and smoked in very much the way we cure and smoke ham. One interesting difference is that the meat is rubbed thoroughly with a dry cure containing 1 cup brandy per 8 lb. salt before it is brined. After smoking, it is aged for 2 mo. before it is baked and served hot. After 4 mo. aging it is eaten without cooking.

NATAL PLUMS

This spiny, tropical, evergreen shrub has 1-in., red fruits that taste like cranberries. To make jelly, wash fruits and slice or crush them. Measure 4 cups into a kettle and add 2 cups water. Bring to a boil, then simmer for 20 min. Extract juice, measure it, and mix with an equal quantity of sugar. Boil to jelly stage. Ladle into hot, sterilized glasses and seal. Each cup of juice yields approximately one 8-oz. glass.

NECTARINES

The only difference between nectarines and their close kin, peaches, is that they have a smooth, fuzzless skin, are a little smaller, and taste a wee bit different. Whatever you do with peaches you can do with nectarines.

NEW ZEALAND SPINACH

This is not a true spinach but is used as a substitute for it. Unlike spinach, which grows in cold weather, New Zealand spinach grows in warm weather. Pick off young leaves and process like greens.

OATS

If you like oatmeal and rolled oats, you had better let the milling industry supply them for you. But oat flour is a simple matter. Thresh and winnow the oats as directed in Chapter 7, and put the grain through a grist mill to make coarse flour.

OKRA

Also called gumbo, okra produces odd, pointed, mucilaginous pods that are boiled or fried or used in stews or soups. Although the pods may grow to 9 in. long, they should be harvested when they are only 3 to 4 in. long.

FREEZING

Cut off stems at the base of the pods, being careful not to open the pods. Follow standard freezing procedure. Blanch for 3 min. Store for 10–12 mo. To serve, boil for 6–8 min.

CANNING

Cut off stems at the base. Place pods in boiling water and boil for 1 min. Pack in jars to 1/2 in. of top. Add 1/2 tsp. salt to pints; 1 tsp. to quarts. Cover with boiling water to 1/2 in. of top. Seal. Process in pressure canner at 10 lb. pressure. Pints for 25 min.; quarts for 40 min.

PICKLED OKRA

Makes 4 pints

- 2 lb. okra (3-in. pods)
- 12 celery leaves
- 4 sprigs dill weed
- 4 garlic cloves
- 2 cups water
- 2 cups white vinegar
- 2 tbs. salt

Cut off stems and pack pods into sterilized jars. Add 3 celery leaves, 1 dill sprig, and 1 peeled garlic clove to each jar. Bring water, vinegar, and salt to a boil and pour over the okra. Seal. Let stand for at least 1 mo. before using.

OLIVES

Preserving olives is a tricky job, reasonably difficult and potentially very dangerous if you don't do it right. In view of all this, we turn to the following directions from a California Agricultural Extension Service booklet titled "Home Pickling of Olives."

GREEN-RIPE OLIVES

This process produces straw-yellow to green or brown olives. Choose fruit that is green, straw-colored, or cherry red. Mission olives are particularly desirable because they can be stored in a freezer for 12 mo. whereas other varieties soften badly.

Fill a glass or stoneware container (never a metal container) with a measured gallonage of water and stir in 4 level tbs. flake lye (household lye) per gallon. Let the solution cool to a maximum of 70°. Then add washed and picked-over olives and keep them submerged with a cloth, piece of wood, or a plate. Stir with a wooden spoon every 2 or 3 hr. for 10 to 12 hr.—until the lye reaches the pits. To judge the amount of penetration, cut open large olives now and then. The lye changes the flesh to a yellowish-green color.

If the lye does not reach the pits by your bedtime, discard the solution and cover the olives with water. Then, the next morning, put the olives in a solution of 2 tbs. lye per gallon of water and let stand until the solution reaches the pits. However, if this fails to happen within 30 hr., discard the solution and replace it with a slightly stronger solution of 3 tbs. lye per gallon of water.

When the lye finally reaches the pits—however long that may take—discard the solution, rinse the olives twice in cold water, and cover them with cold water. Change the water 4 times daily until the olives no longer taste of lye—from 4 to 8 days.

During this entire lye-solution and water-soaking process, keep the olives weighted down in the liquid.

Then prepare a solution of 6 1/2 tbs. salt per gallon of water, pour it over the olives, and let them stand for 48 hr. The olives are then ready for consumption, but they can be stored in the refrigerator or a strong brine solution for only a couple of weeks.

To prepare the olives for long-term storage, leave them in the solution of 6 1/2 tbs. salt to 1 gal. water and boil them for 10 min. Then throw out the brine and cool the olives completely in cold water. Drain them and pack in rigid freezer containers, and freeze. They can be stored for as much as 12 mo. Before serving, allow the olives to thaw thoroughly.

SPANISH-STYLE GREEN OLIVES

These olives have a green skin, light flesh, and light brownish-buff pit. You can use any variety, though Sevillano and Manzanilla are preferred. Olives are ready for preserving when green to straw-yellow in color. Discard defective fruit and take pains not to bruise any fruit because the marks are accentuated by the pickling.

As soon as you have picked over and washed the olives, place them in a cold lye solution. For Sevillanos, use 3 1/2 to 4 tbs. lye per gallon of water. For Manzanilla and Mission varieties, use 4 1/2 to 5 1/2 tbs. lye. Keep the olives below the surface. Let the lye penetrate about three-quarters of the way to the pits; then pour off the solution and put the olives in cold water. Change the water every 4 hr. for the next 30 hr.

When the lye taste is gone, pack the olives as quickly as possible into containers of at least 1 qt. capacity, and cover them with brine. Use 15 1/4 tbs. salt per gallon of water for Sevillanos; 26 tbs. salt per gallon for Manzanillas and Missions. Cover with caps but don't tighten these all the way. Store the containers in a dark place at 70° to 90°.

During the first 4 or 5 days of storage, fermentation should be quite active. Check the containers frequently and immediately replace any brine that is lost. When fermentation becomes less active, tighten caps, but continue to check the containers and replace any lost brine at once.

If you wish to make Manzanilla and Mission olives more acid, add sugar to the brine after active fermentation has ceased. Use it at the rate of 1 1/2 tsp. per gallon of brine. Also to make the olives more acid, you may add 3/4 cup dill-pickle or sauerkraut brine to each gallon of the olive brine.

When the olives have attained the desired flavor, fill the containers to the rim with brine; wipe off the rims; and cap the containers tightly. Store in a cool, dark place. They will keep for a year or more.

OLIVE OIL

Olives may be harvested for oil from the time they ripen until they turn black. Oil content of blackened fruit is better than that of green fruit, but the oil is likely not to be of such good quality.

Green fruit should be allowed to dry for

a week or two in trays in a dry, airy place or it can be dried in an oven at 120° for about 48 hr. Turn the fruit daily during the outdoor drying process.

To press out oil, use a small cider mill. First crush the olives (but not heavily enough to crack the pits), then press out the juice consisting of oil and water. Strain this through muslin or several layers of cheesecloth and allow it to stand. When the oil rises to the top, siphon it off.

Clarification of the oil is accomplished by letting it stand in glass or crockery containers until the sediment settles, then siphoning off the clear oil through cheesecloth. This should be done 3 or 4 times over a period of several months. The clear oil is then poured into sterilized bottles and capped tightly.

ONIONS

DRY STORAGE

Not all onions are good keepers. Those that do well include Downing's Yellow Globe, Early Yellow Globe, Ebenezer, Red Wethersfield, Southport White Globe, White Portugal, and Yellow Globe Danvers.

If you grow your own onions (something we didn't do for a long time, but something we wouldn't miss doing now), wait until about half of the tops have tipped over; then tip over the rest by hand, and let the bulbs ripen for 3 or 4 days. Then pull them up, cut off tops 1 in. above the bulbs, and continue to dry the bulbs (don't wash off the soil) in a shady, airy place until the necks are completely dry. Any onions taking an unusually long time to dry or having damaged necks should be used promptly. Store the rest in mesh bags or slatted crates in a dry, well-ventilated storage room. At the ideal temperature of close to 32°, the onions will keep for up to 8 mo.

FREEZING

Peel and chop into tiny pieces. Pack in small polyethylene bags or rigid containers. Freeze at once. Use for seasoning. Store for 10–12 mo.

DRYING

Use pungent varieties. Peel. Cut into 1/8- to 1/4-in. slices. Spread in a shallow layer on trays and dry in a 140° oven until brittle—about 6 to 10 hr. Package slices whole in tight containers or crush to a powder and pack in bottles.

PICKLED ONIONS

Makes 6 pints

8 lb. small white onions
1 cup salt
2 qt. white vinegar
2 cups sugar
3 tbs. mustard seed
3 tbs. black peppercorns
3 tbs. grated horse radish
6 small red pepper pods
6 bay leaves

Let onions stand for 2 min. in boiling water, then put in cold water and peel. Put in bowl, sprinkle with the salt and cover with cold water. Let stand overnight. Then drain and rinse well under running water. Boil vinegar, sugar, mustard seed, pepper, and horse radish for 2 min. Add onions and bring to a boil. Pack onions into hot, sterilized jars. Add 1 pepper pod and 1 bay leaf to each jar. Cover onions with boiling sirup. Seal.

OPOSSUM

The opossum may be a very ugly animal but he has very tender, fine-grained meat which is well larded with fat. Remove the entrails and head as soon as possible after shooting. Then let the carcass hang in a cool place for 24 to 48 hr. before skinning and removing feet (it's a hard job otherwise). Cut into serving pieces and remove as much fat as possible. Wrap all pieces together, tightly, in aluminum foil, and freeze. Store for 4–6 mo.

ORANGES

The following sage advice comes from the California Agricultural Extension Service:

> Citrus fruits can be left on the tree for considerable periods of time without deterioration. Whenever possible this is the best method of storage. They are, however, subject to damage by wind and frost and eventually they become overripe and unpalatable. For this reason the fruit is sometimes picked and stored for future use. The fruit, if uninjured, is resistant to decay and can be kept for several weeks if properly handled and stored. The longer the fruit remains on the tree after maturity, the shorter the time it can be kept in storage.
>
> Citrus fruits which are to be stored must be harvested and handled with the greatest of care. Any break in the rind will open the way for decay organisms. Always wear soft gloves when picking or handling the fruit, for it is almost impossible to avoid cutting the rind with fingernails when bare hands are used. Do not pull the fruit from the tree, but clip the stem with a close, smooth cut. Rough or long stems puncture other fruits during handling. Use clippers with care, for clipper cuts and bruises are a frequent cause of decay. Avoid scratching the fruit on thorns or dead brush and do not drop the fruit when putting it into bags or boxes. When it is being transferred from one container to another, pour it carefully. See that all the containers are free from twigs and gravel, which might damage the rind. Pick the fruit when it is thoroughly dry. Wet fruit is more easily damaged than dry fruit.
>
> Citrus fruits keep best in cool rather than cold storage. Temperatures in the neighborhood of 60° F are satisfactory. Choose a place where the temperature is or can be held at a fairly uniform level.
>
> Do not attempt to store fruit that has been injured; remove any that shows decay during storage. A few bad fruits will hasten deterioration in the whole lot. The less citrus fruit is handled the better it will store.
>
> Place the fruit in convenient containers (clean wooden boxes are usually best) and get it into storage as soon as possible after picking and with a minimum of handling. Fruits may be stored bare or wrapped individually. Wrapping tends to isolate decaying fruit and prevents the withering that may occur if the air is too dry.

FREEZING

Wash, peel, and section oranges and pack in containers with 30 percent sugar sirup. Some of the sirup may be made of orange juice. Store for 6 mo.

JUICE

Extract juice from fruit with a reamer, not a press. If juice is to be canned, follow procedure outlined in Chapter 9. If juice is to be frozen, pour it into glass freezer jars (plastic may affect flavor). Seal and freeze. Store for 3–4 mo.

Do not make juice from navel oranges. Valencias make the best juice for freezing.

MARMALADE

Without added pectin
Makes 7 8-oz. glasses

4 oranges
4 lemons
water
salt
sugar

Remove peel from oranges and lemons, lay it flat, and pare off and discard about half of the white part. Then slice the peel very thin. Cover with cold water and boil till tender. If the peel tastes very bitter, change the water several times. Chop fruit pulp and remove seeds and rag. Combine pulp and peel and measure (pack the fruit down firmly). To each cup fruit add 2 cups water and 1/8 tsp. salt. Boil hard for 25 min., then measure and add an equal measure of sugar. Boil to jelly stage. Skim and ladle into hot, sterilized glasses. Seal.

MARMALADE

With added pectin
Makes 7 8-oz. glasses

3 oranges
2 lemons
1 1/2 cups water
5 cups sugar
1/2 bottle liquid pectin

Remove orange and lemon peels, lay them flat and pare off and discard about half of the white part. Slice peel into very thin slices. Cover with 1 1/2 cups water, bring to a boil, and simmer for 20 min. Chop the pulp and discard seeds and rag. Add the pulp to undrained peel and simmer for 10 min. Measure 3 cups fruit into a kettle and mix in sugar. Follow standard procedure for making jam with liquid pectin.

CANDIED PEEL

1 or 2 oranges
1 cup sugar
1/2 cup water

Peel oranges. Cut peel into lengthwise strips, cover with water, bring to a boil, and simmer till soft. Drain and lay flat. Pare off white part of peel and discard. Slice remaining peel into narrow strips. Bring sugar and water to a boil, add peel, and simmer until sirup reaches 230°. At this point the peel should be clear. Remove it from sirup, spread on a plate, and roll in confectioner's sugar. When cold, store in glass containers.

ORANGE ICE

3 cups orange juice
juice of 1 lemon
1/2 cup water
2 cups sugar
pinch salt

Heat water and 1 cup orange juice with sugar and salt until sugar dissolves. Chill completely. Combine with remaining orange juice and lemon juice. Pour into a pan and freeze firm.

PECTIN

Use peels of oranges from which juice has been extracted. Grate off the orange skin and scrape off pulp clinging to the white peel. Cut into small strips and place 4 packed cups in a kettle with 2 qt. water and 1 tbs. tartaric acid (from the drugstore). Let stand for 2 hr. Measure depth of mixture and then boil until depth is reduced 50 percent. Strain into a bowl through four or five thicknesses of cheesecloth. Then repeat process 2 more times.

Mix the 3 extractions together, bring to a boil, and pour into hot, clean Mason jars to 1/2 in. of top. Seal. Process in simmering water to cover for 30 min.

This pectin can be used instead of commercial liquid pectin at the rate of 1/4 cup to each cup of fruit juice.

OREGANO

Oregano, or pot marjoram, is a perennial herb with flavorful leaves used for seasoning poultry, fish, soups, salads, sauces, and tomato juice. It is a special favorite of Italian and Spanish cooks. Dry the leaves in a warm, airy place before the flowers form. Then crush the leaves and store them in tight containers.

OREGON GRAPES

Oregon grape is an evergreen shrub with hollylike leaves. It is grown for ornament and found in the woods in our temperate climates. It bears clusters of small, oval, blue berries which should not be eaten fresh because they have a laxative effect. But they make a good jelly, which should also be eaten in moderation.

Crush berries and add just enough water to prevent scorching. Bring to a boil and simmer for 10 min. Extract juice. Measure into a kettle and add an equal amount of sugar. Follow standard procedure for making jelly without added pectin.

OYSTERS

One of the early Virginia explorers noted in his chronicles: "I have seen [oysters] some thirteen inches long; the salvages used to boil oysters and mussels together and with the broath they make a good spoone meal, thickened with the flower of their wheat; and it is a great thrift and husbandry with them to hang oysters upon great strings being shauled [shelled] and dried in the smoake, thereby to preserve them all the year."

For how to smoke oysters yourself, see Chapter 4. Or try this recipe from Mrs. Clifford Smith of Corvallis, Oregon:

Select not-too-large oysters of even size. Be sure they are fresh. Shell them; then wash and free the meats from all shell particles. Boil together 1 gal. water and 2/3 cup dairy salt; reduce to a simmer and drop in oysters for 4 to 5 min. Then drain and dry.

Place the oysters on wire racks in the smokehouse and smoke for 3 or 4 hr. Don't let them dry out. Then pack the oysters solidly in half-pint jars to 1/2 in. of top. Cover with olive oil. Process in pressure canner at 10 lb. pressure for 100 min.

FREEZING

Wash shells and cut them open with an oyster knife over a pan in order to save the juice. Wash meats well in salty water to remove sand. Pack into rigid plastic containers or glass freezer containers and cover with the strained juice or a cold brine made in the proportions of 1 gal. water to 1 tbs. salt. Leave 1/2 in. head space. Freeze. Store for 6 mo. Ready to serve when thawed.

CANNING

Prepare as above and place washed meats in clean Mason jars—1 cup meats in half pints; 2 cups in pints. Fill jars with strained oyster juice and/or a brine made of 1 qt. water and 2 tbs. salt. Place jars in a kettle in water to within 2 in. of rims, cover kettle, and bring water to a boil. Boil for 5 min. Then seal jars and process in a pressure canner at 10 lb. pressure. Half pints for 35 min.; pints for 50 min.

PICKLED OYSTERS

Shell oysters and simmer them in their own liquor very gently for 10 min. Drain off liquor; and while oysters are cooling, mix it with an equal quantity of white vinegar, a sprinkle of mace, 1 strip lemon peel, and about 6 peppercorns. Boil for 5 min. and let cool. Place oysters carefully in sterilized Mason jars and cover them with the pickle to 1/4 in. of top. To keep oysters from rising out of the pickle, crush a little plastic film into the top of the jar. Seal with two-piece lids. Store for 1 mo. in a cool, dark place or for a longer period in the refrigerator.

PAPAYAS

Papayas resemble a yellow melon with orange flesh, but they have a flavor and texture of their own; and as sources of Vitamins A and C, they are excellent. But unfortunately there are not too many ways to put them up.

CANNING

Wash, cut, and seed firm, ripe papayas. Slice into a cold 40 percent sirup to which the juice of 1 lemon or lime has been added. Bring slowly to a boil and cook for 2 min. Pack in jars and cover with sirup to 1/2 in. of top. Seal. Process in boiling water. Pints for 15 min; quarts for 20 min.

Green papayas can also be canned. Wash and peel fruit and boil for 4 min. in water. Drain. Combine 3 1/2 cups sugar, 1 cup vinegar, 1 cup water, 1/2 oz. ginger root, 2 tbs. ground cinnamon, and boil till sugar is dissolved. Strain and pour over the papayas and bring to a boil. Then pack fruit into jars and cover with sirup to 1/2 in. of top. Seal. Process in boiling water. Pints for 15 min.; quarts for 20 min.

JAM

Makes 4 8-oz. glasses

3 lb. papayas
3 cups sugar
1/2 cup orange juice

Wash and halve ripe papayas; scoop out seeds; then scoop out flesh and mash. Measure 3 cups into a kettle. Add sugar and orange juice. Simmer, stirring occasionally, until mixture meets a jam test. Ladle into hot, sterilized glasses and seal.

PRESERVES

Wash, peel, and cut fully ripe papayas into uniform pieces. The seeds need not be removed. Weigh. Sprinkle 1 lb. sugar over every pound of fruit. Let stand overnight. If enough liquid is not drawn out of the fruit to cover them, add a little water. Bring to a boil and boil till fruit is clear—about 10 min. Cover and let stand overnight. In the morning, boil till sirup is thick. Ladle into hot, sterilized jars, cover with hot sirup and seal.

SHERBET

1 1/2 lb. papayas
3 tbs. lemon juice
1/2 cup orange juice
1 cup sugar
1 1/2 cups milk

Wash and halve ripe papayas; scoop out seeds; then scoop out flesh and mash it or purée in a blender. Measure out 1 1/2 cups. Add fruit juices. Dissolve sugar in milk and gradually add fruit. Pour into pans or rigid plastic containers and put in freezer. Stir every 30 min. until mixture is frozen hard.

PARSLEY

Probably the most widely used of all herbs, parsley is a pretty biennial plant which grows everywhere. To dry the leaves for seasoning, cut young stems from the plant, wash quickly under cold water to remove dust and dirt and shake as dry as possible. Place on a piece of screen cloth in a 400° oven and heat for 5 min. or a little longer until the leaves are very crisp. Turn the stems once during heating. When cool, rub the leaves through a coarse sieve and store them in tight containers.

Green sprigs of parsley can also be frozen without blanching. Pack them in tightly sealed containers or polyethylene bags that hold just enough for each use.

An unusual variety of parsley named Hamburg develops long, 2-in.-thick roots resembling parsnips. These may be cooked like parsnips or cut into small pieces to flavor stew, soups, etc. To store, dig up the roots in the fall, cut off the tops, and place the roots, without washing, in boxes filled with sand. Store in a cold root cellar.

PARSNIPS

The big, long, carrotlike white roots of the parsnip have a distinctive flavor and are delicious when boiled or fried. In cold climates, they can be left outdoors all winter and dug up as needed, because they are improved by freezing. They can also be dug in the fall and stored, surrounded by soil, in boxes stored in a cold root cellar.

In warm climates, however, parsnips should be frozen. Use young, tender roots without woody cores. Cut off tops, peel, and cut lengthwise in 1/2-in. slices. Follow standard freezing procedure. Blanch for 2 min. Store for 10–12 mo. To serve, cook for 12–15 min.

PARTRIDGE

Handle and freeze like pheasant.

PASSION FRUIT

This tropical evergreen vine, widely grown in Hawaii, produces 2-in.-diameter purple or yellow fruits which are generally eaten fresh but which may also be cooked or preserved.

JELLY

Makes 3 8-oz. glasses

3 lb. passion fruit
3 cups sugar
1 cup water
1/2 bottle liquid pectin

Wash and cut fruits in half; scoop out pulp and seeds and mash in a sieve to extract juice. Strain through cheesecloth. Measure out 1/2 cup and set aside. Bring sugar and water to a boil and boil hard for 1 min. Remove from heat and immediately stir in pectin. Then pour in fruit juice and mix well. Skim. Ladle into hot, sterilized glasses and seal.

JUICE

Cut washed fruits in half and scoop out pulp and seeds. Place in a jelly bag and squeeze out juice. Mix 1 cup sugar with 3 1/2 cups juice. Pour into glass freezer jars, seal and freeze. Store for 3–4 mo.

SHERBET

1 1/2 cups sugar
2 cups water
pinch of salt
2 tbs. light corn sirup
1 cup passion fruit juice
2 tbs. lemon juice
2 beaten egg whites

Cook sugar, water, salt, and corn sirup together until sirup spins a thread. Cool. Add fruit juices, pour into pan, and place in refrigerator or freezer. When mixture is half-frozen, fold beaten egg whites into it and freeze solid.

PAWPAWS

Pawpaws, or papaws, are fairly small deciduous trees which grow in the East and Middle West. They bear misshapen, brown fruits, considerably longer than wide, with a soft, yellow, banana-flavored flesh. They are picked as soon as they begin to mellow and are allowed to continue ripening in a cool place.

The Indiana nurseryman from whom we bought our last pawpaw tree (which mysteriously disappeared soon after planting) drew a blank when we asked him for suggestions about preserving the fruits, but he was kind enough to refer us to J. C. McDaniel, Associate Professor of Horticulture at the University of Illinois.

The professor wrote: "While I've had no experience in culinary preparation of papaw, I did have one accidental experiment with it years ago. A hard-mature fruit was left in the pocket of a leather jacket hanging in a closet. After about a month I found it dried and quite edible."

So there is one way you can put up pawpaws. Another way is to wash, peel, and seed the fruits. Pack the pulp in rigid plastic containers and freeze. Store for 4–6 mo. When thawed, you will find the fruit very soft. Eat it promptly.

PEACHES

To peel peaches that are to be put up, dip them in boiling water for about 30 sec. or until the skins feel loose when you pinch them; cool in cold water, and zip off the skins.

Varieties especially good for processing are Elberta, Blake, Golden Jubilee, Halehaven, J. H. Hale, Redhaven, Redskin, Rio Oso Gem, and Southland.

FREEZING

Peaches to be frozen can be peeled as above, but since the boiling water softens the flesh slightly, it is better to peel them without dipping. Pit the fruit and slice or leave in halves. Place in containers partially filled with a 30 or 40 percent sugar sirup to which ascorbic acid has been added. (Use 4 tsp. ascorbic acid per 4 cups sirup.) Cover with sirup, seal, and freeze.

Peaches to be used in pies and for other cooked dishes are best packed in sugar to which ascorbic acid has been added. Use 1 lb. sugar and 2 tsp. ascorbic acid per 4 or 5 lb. fruit.

Crushed or puréed peaches are also packed with sugar to which ascorbic acid has been added. Use sugar and ascorbic acid at the rate given above or sweeten to taste.

All frozen peaches can be stored for 10–12 mo.

CANNING

Raw pack. Peel and pit peaches; slice or leave in halves. As fruit is prepared, drop it into a solution of 1 gal. water and 2 tbs. salt and 2 tbs. vinegar to prevent darkening. Then drain and pack in jars to 1/2 in. of top. Cover with boiling 40 percent sugar sirup to 1/2 in. of top. Seal. Process in boiling water. Pints for 25 min.; quarts for 30 min.

Hot pack. Prepare peaches as above and heat through in 40 percent sugar sirup. Pack in jars and cover with sirup to 1/2 in. of top. Seal. Process in boiling water. Pints for 20 min.; quarts for 25 min.

DRYING

Use peaches just ripe enough to eat. Peel, pit, and cut in halves, quarters, or slices. Drop fruit as it is prepared into salt and vinegar solution described above. Then sulfur for 4 hr. Place in a single layer in trays, pit side up, and dry in the oven till leathery. Start at 125° and gradually increase to 175°.

JUICE

Wash and stem firm, ripe fruit. Place in boiling water 1 in. deep and boil till soft. Pit and press through a food mill. Strain through cheesecloth or a jelly bag to extract juice. Make a sirup of 1 cup sugar and 4 cups water and mix with the fruit juice in equal parts. Then follow standard procedure for making fruit juice.

ICE CREAM

See *Milk.*

JAM

Without added pectin
Makes 3 8-oz. glasses

1 lb. peaches
1 1/2 cups sugar

Wash and peel peaches, pit and crush in a kettle. Mix in sugar and let stand for 3 hr. Heat slowly until sugar dissolves. Then follow standard procedure for making jam without added pectin.

JAM

With added pectin
Makes 10 8-oz. glasses

3 lb. peaches
1/4 cup lemon juice
7 1/2 cups sugar
1/2 bottle liquid pectin

Wash, peel, pit, and crush peaches. Measure 4 cups into a kettle. Add lemon juice and sugar. Follow standard procedure for making jam with liquid pectin.

JELLY

Makes 10 8-oz. glasses

3 1/2 lb. peaches
1/2 cup water
1/4 cup lemon juice
7 1/2 cups sugar
1 bottle liquid pectin

Wash, peel, pit, and crush peaches. Add water, bring to a boil, and simmer for 5 min. Extract juice. Measure 3 1/2 cups into a kettle and add lemon juice and sugar. Follow standard procedure for making jelly with liquid pectin.

PRESERVES

Makes 4 half-pints

1 1/2 lb. peaches
2 cups sugar

Use small peaches if they are to be left whole. Peaches can also be halved or quartered. Wash, peel, and pit (except whole peaches). Mix with sugar and let stand for 12 hr. Then bring slowly to a boil and cook till sirup is rather thick. Put fruit in hot, sterilized jars. Cover with sirup and seal.

PICKLED PEACHES

Makes 7 pints

7 lb. small peaches
whole cloves
6 pieces stick cinnamon
8 cups sugar
1 qt. cider vinegar

Leave peaches whole. Wash, peel, and stick 3 or 4 cloves into each one. Boil cinnamon, sugar, and vinegar together for 2 min.; add half the peaches and cook gently till they are soft. Pack in hot, sterilized jars; cover with sirup and seal. Then cook remaining half of peaches in remaining sirup and package.

BRANDIED PEACHES

Wash and peel peaches. Cut in half and pit, or leave small peaches whole. Mix 1 lb. sugar with 1 lb. fruit. Place in jars or a crock, cover with brandy, and close lid. Let stand for a fortnight or longer before using.

CANDIED PEACHES

See Chapter 10. Fruits should be peeled, pitted, and cut in half.

PEANUTS

Peanuts are annual vegetables related to peas. They grow about 18 in. tall, produce clusters of seed pods underground. The nuts are harvested in the fall by digging up the entire plant when the leaves turn yellow and the seeds in the pods are pink or red. Shake off as much dirt as possible and spread out the plants to dry in a warm, airy but not sunny place for 2 or 3 wk. The nuts can then be picked off and stored in a cool, dark place.

Shelled peanuts are canned according to directions in Chapter 8.

Peanuts are also occasionally canned in their shells. Use nuts directly from the garden; do not cure them. Wash thoroughly; place in a kettle containing 1 tbs. salt per quart of water. Boil for 15 min. Then pack the nuts in Mason jars to 1 in. of top. Pack firmly but do not crush. Cover with the cooking water. Seal. Process in a pressure canner at 10 lb. pressure. Pints for 40 min.; quarts for 50 min.

PEANUT BUTTER

Roast cured, cleaned nuts in their shells at 300° for 45 min. Stir frequently. Shell, rub off skins, and remove the small meaty germs at the ends of the nuts. Put through a meat grinder, using the finest plate, until the butter is uniformly smooth. Mix in salt to taste. Then pack in sterilized jars and seal.

PEARL MILLET

Millet was one of the first grains consumed by man; and in other parts of the world it is still used in this way very extensively. However, the millet plant from which this food is derived is not grown in the United States.

The best millet we grow for food purposes is pearl millet. Found only in warm climates, it is a tall, clump-forming grass topped by dense, round seed heads up to 1 ft. long. Thresh, winnow, and mill the grain as in Chapter 7.

PEARS

DRY STORAGE

Pears should be picked before they are ripe—when the green skin begins to yellow and you can pull the fruits from the tree with a slight tug. If they are to be eaten or processed soon thereafter, allow them to ripen out of the sun at a temperature of less than 70°.

For long-term storage, on the other hand, put the pears directly into the fresh-food section of your refrigerator. Some varieties can be stored for several months. Those that do particularly well are Bartlett, Beur-Bosc, Anjou, and Magness.

CANNING

Pear varieties especially good for processing include Baldwin, Bartlett, Gorham, Kieffer, Maxine, Moonglow, and Orient.

Raw pack. Wash, peel, halve, and core pears. Until you're ready to process them, place them in 1 gal. water to which 2 tbs. salt and 2 tbs. vinegar have been added. Then drain and pack in jars to 1/2 in. of top. Cover with hot 40 percent sugar sirup to 1/2 in. of top. Seal. Process in boiling water. Pints for 25 min.; quarts for 30 min.

Hot pack. Prepare as above. Add pears to hot 40 percent sugar sirup, bring to a boil, and cook for 5 min. Pack in jars and cover with sirup. Leave 1/2 in. head space. Seal. Process in boiling water. Pints for 20 min.; quarts for 25 min.

FREEZING

It is much better to can pears than to freeze them. But if you insist—wash, peel, halve, and core the fruit. Heat for 90 sec. in boiling 40 percent sugar sirup to which ascorbic acid has been added. (Use 4 tsp. ascorbic acid per 4 cups sirup.) Drain and chill. Chill sirup. Pack pears in containers, cover with sirup, seal, and freeze. Store for 10–12 mo.

DRYING

Wash, peel, and core pears and cut in slices. Put them in salt and vinegar solution till ready to process. Then sulfur for 2 hr. Spread in trays in a single layer and dry in oven until springy. Start at 130° and gradually increase to 150°. Package in tight containers and store in a cool, dry, dark place.

JAM

Makes 10 8-oz. glasses

3 lb. pears
1/4 cup lemon juice
7 1/2 cups sugar
1/2 bottle liquid pectin

For some reason pear jam generally gets overlooked. This is a shame. It makes a meltingly delicious product. Wash, peel, and core pears. Chop fine. Measure 4 cups into a kettle with the lemon juice and sugar. Follow standard procedure for making jam with liquid pectin.

PRESERVES

Makes 9 half pints

4 lb. pears
2 lemons
6 cups sugar
4 cups water

Pears should be hard-ripe. Wash, peel, and core them, and cut in halves or quarters. If you are using Kieffer pears, boil in water until they can be pierced with a fork —about 20 min. Slice lemons, combine with sugar and water; bring to a boil. Add pears and cook rapidly until fruit is clear and sirup meets a jelly test. Then pour mixture into a shallow dish and let the fruit plump overnight. In the morning, bring sirup to a boil; pack fruit into hot Mason jars; and cover with sirup to 1/4 in. of top. Seal. Process in boiling water for 15 min.

PICKLED PEARS

Makes 7–8 pints

2 tbs. whole cloves
2 tbs. whole allspice
8 cups sugar
1 qt. white vinegar
2 cups water
8 pieces stick cinnamon
8 lb. Seckel pears

Tie cloves and allspice in a bag and combine with sugar, vinegar, water, and cinnamon. Bring to a boil and simmer for 30 min. Wash, peel, and remove blossom ends of pears but leave stems on. Place in 1 gal. water, 2 tbs. salt, and 2 tbs. vinegar until ready to process. Then drain, add to sirup, and continue simmering for 20 min. Pack pears in hot jars, add a piece of cinnamon to each jar, and cover with boiling sirup. Leave 1/2-in. head space. Seal. Process in boiling water for 20 min.

CANDIED PEARS

See Chapter 10. Prick skins with a sharp fork, cut fruits in half, and remove cores.

PEAS

FREEZING

Varieties especially good for freezing include Frosty, Laxton's Progress, Lincoln, and Wando. Don't waste any time getting peas from the garden into the freezer. Follow standard freezing procedure. Shell and blanch for 1 min. Store for 10–12 mo. To serve, cook for 3–5 min.

Edible podded peas, often called sugar peas, should be picked when the seeds begin to plump up—before they are fat. Remove stems, wash pods, and blanch for 90 sec. Store for 10–12 mo. To serve, cook for 4–5 min.

CANNING

Raw pack. Pick peas just before canning. Shell and wash. Pack loosely in jars to 1 in. of top. Add 1/2 tsp. salt to pints; 1 tsp. to quarts. Cover with boiling water to 1 in. of top. Seal. Process in pressure canner at 10 lb. pressure for 40 min.

Hot pack. Shell, wash, cover with boiling water, and bring to a boil. Drain and pack loosely in jars to 1 in. of top. Add 1/2 tsp. salt to pints; 1 tsp. to quarts. Cover with cooking water. Seal. Process in pressure canner at 10 lb. pressure for 40 min.

DRYING

Pick peas when young and tender. Shell and blanch in boiling water for 3 min. Drain thoroughly. Dry in 150° oven until brittle—about 6 to 10 hr. Seal in polyethylene bags or rigid containers and store in a cool, dark place.

PECANS

The only thing wrong about pecans—and it is very wrong—is the way northerners mispronunce the name. The proper pronunciation is "pee-cons" with the accent on the second syllable.

Otherwise, these are beautiful, beautiful nuts. When we visit our southern cousins, someone sooner or later brings out a sack of pecans and everybody sits around cracking and eating until there isn't inner space for another nut. And still the cracking and eating continue.

It's easy to tell when pecans are ready for harvesting because the green husks turn brown and open at the tips. You can leave the nuts on the tree to fall of their own accord, or if you're impatient, you can knock them down with bamboo poles.

If the weather has been damp or humid, spread the nuts out in a dry, airy place to dry completely before you bundle them in sacks. In dry weather, however, bagging can be done at once.

We have never had any trouble storing unshelled pecans at normal room temperature for 4–6 mo.; but maybe we have been plain lucky, because food technologists say that a 40° temperature is needed for under 6-mo. storage and 32° for under 12-mo. storage.

Shelled pecans, in our experience, can also be stored at room temperature for a number of months if kept in tight polyethylene bags or—in view of the mice problem—in canisters. But there is no question that storage life will be extended greatly if you freeze the kernels or can them according to directions in Chapter 8.

PEPPERS, CHILE

Whether you call them chile peppers or hot peppers, and no matter how you put them up, the thin, tough skins found on many varieties must be removed first. Of the various ways of doing this, the following are as easy and safe as any:

Wash peppers and put them in a 450° oven for 6 min. Remove from heat and wrap in a hot, wet towel to steam for 15 min. Then peel off the skins.

Blister the skins over a flame, turning often to prevent scorching. Then wrap peppers in a wet towel for 10 min. before peeling.

Wash and dry peppers well. Prick skins with a fork to keep them from exploding. Lower into hot paraffin until skin blisters and whitens, about 4 min. Place in ice-cold water. Then peel. The skin comes off with the wax.

FREEZING

Use green peppers. Peel and cut out stems and seeds. Then flatten the whole chiles to remove air, and fold once. Pack in rigid plastic containers. If you put squares of wax paper between the chiles, they will separate more easily when you thaw them. Put in freezer. Store for 10–12 mo.

CANNING

Use green peppers. Peel and remove stems and seeds. Flatten the whole chiles and fold once. Pack tightly in half-pint jars to 1/2 in. of top. Add 1/2 tsp. salt. Don't add liquid. Place jars, uncovered, in a kettle with hot water about 2 in. below the jar rims. After boiling for 12 min., press chiles down with a spoon. This will leave extra empty space in the jars, so use the chiles in one jar to fill the other jars to 1/2 in. of top. Continue boiling for 8 min. (a total of 20). Seal jars and process in a pressure canner at 10 lb. pressure for 20 min.

DRYING

Green chiles. Wash, peel, slit, and remove seeds and stems. Open flat, spread in a single layer in trays, and sun-dry until peppers are crisp, brittle, and medium green—about 2 days.

A better method is to prepare peppers as above. Then put them in a strainer over boiling water, cover, and steam for 10 min. Spread in single layers in trays and dry in a 150° oven until crisp and brittle.

Red chiles. Use mature peppers. Wash. String whole peppers together with needle and cord and hang them outdoors in large bunches to sun-dry. When fully dry, the pods will be shrunken, flexible, and dark red.

PICKLED CHILES

Makes 4 pints

2 lb. peppers
9 cups cider vinegar

Peel and remove stems and seeds from peppers. Cut large peppers into pieces or leave whole. Very small peppers can be left with stems and seeds. Pack tightly into jars. Heat vinegar to 160° (simmering) and pour over peppers, covering them well. Put on caps loosely. Place jars out of the sun at a temperature of 60° to 75°. The vinegar will form bubbles of escaping gas. When bubbling stops (this may take several days to 2 wk.), drain off vinegar and cover peppers with fresh vinegar to the jar rim. Add salt to taste. Seal and store. If you prefer slightly oily pickles, fill jars to 3/4 in. of top with vinegar and fill the rest of the way with olive oil. The peppers will be coated with oil as you take them out of the jars.

JELLY

Makes 8 8-oz. glasses

6–7 green sweet peppers
12 red chile peppers
6 1/2 cups sugar
1 1/2 cups cider vinegar
1 bottle liquid pectin

Wash and cut green peppers in half; remove stems and seeds. Peel chile peppers and remove stems and seeds. Grind green and chile peppers separately. Measure 2 cups tightly packed green peppers and 1/4 cup tightly packed chile peppers into a kettle. (Use juice as well as pulp if pulp is short.) Add sugar and vinegar and bring to a boil; then remove from heat and let cool for 20 min. Stir occasionally. Then return to the range and bring mixture to a rolling boil and boil for 2 min. Remove from heat at once and stir in pectin. Skim and stir for 5 min. Then pour into hot, sterilized glasses and seal.

This jelly is prettier—but not tastier—if you can find red sweet peppers.

PEPPERS, SWEET

FREEZING

Peppers can be frozen without blanching if you want to use them in uncooked foods such as salads. Start with firm, crisp red or green fruits; wash and cut in half. Remove stems and seeds. Then cut flesh into 1/2 in. strips or rings. Pack without head space into rigid containers and freeze.

Pimientos are frozen as above; but after removing stems and seeds and before cutting into strips, you must peel off the skins. To do this, place peppers in a 400° oven for 3 to 4 min., then rinse under cold water to remove charred skin and to cool peppers. Drain; slice as desired; pack in freezer containers, and put in freezer.

If peppers are blanched, they lose crispness but are excellent for cooking. They also pack more easily. Use firm, crisp red or green fruits. Cut in half and remove stems and seeds. You can then freeze the peppers as is or cut them into strips. Follow standard freezing procedure. Blanch halves for 3 min.; strips for 2 min. To serve, cook for 5 to 8 min.

Store all peppers for 10–12 mo.

CANNING

Use small green peppers. Cut in half, remove stems and seeds. Boil for 3 min. Pack into pint jars to 1 in. of top. Add 1 tbs. vinegar and 1/2 tsp. salt. Cover with boiling water to 1 in. of top. Seal. Process in pressure canner at 10 lb. pressure for 35 min.

BRINING

See Chapter 6.

PICKLED PEPPERS

Makes 8 pints

6 lbs. peppers
2 1/2 qt. vinegar
1 1/2 qt. salad oil

Dip washed peppers in boiling water for 1 min. Leave small peppers whole; just remove stems, seeds, and cores. Cut large peppers in half, remove stems, seeds, and cores, and then cut in quarters or eighths. Bring vinegar to a boil and add to peppers. Let stand for 24 hr. or more. Then drain off vinegar and pack peppers in hot, sterilized jars. Meanwhile heat salad oil to about 200°, pour over peppers. Seal jars.

PEPPER-ONION RELISH

Makes 5 half pints

4 or 5 medium sweet red peppers
4 or 5 medium green peppers
4 or 5 large onions
3 cups cider vinegar
1 cup water
2 cups sugar
1 tbs. salt

Wash and cut peppers in half; remove seeds, cores, and white membranes. Peel onions. Put each vegetable separately through a coarse grinder and measure out 2 cups of each. Combine vegetables and cover with 1 cup vinegar and the water. Let stand for 15 min., then drain. Combine sugar, salt, and 2 cups vinegar; add to vegetables. Bring to a boil and boil vigorously for 15 min. Pack into hot, sterilized jars and seal.

PERCH

Several perches are caught in fresh water but the most familiar is the yellow perch, running to about 1 lb. Ocean perches, also called rosefishes, are about five times larger. All are wrapped and frozen in the usual way. Store for 6–9 mo.

PERSIMMONS

We have fiddled with persimmons but can't claim much more. But some years ago we were lucky enough to make contact with Dr. William H. Preston, Jr., of the Department of Agriculture, an expert on the fruit. And recently he introduced us to another expert, his former colleague in the department, Dr. Eugene Griffith. Here is what the latter wrote us about persimmon preservation:

> The customary method for preserving Oriental persimmons in north China is by freezing them. A half century or so ago, when Americans had access to the area, millions were frozen annually outdoors in Peking. The cold winds off the Gobi and from Siberia kept them frozen all winter, or until they were consumed. In Japan, with off-shore breezes and a warmer climate, drying was the customary method for preserving persimmons. And again, millions were dried annually. (I don't mean to imply that the Chinese in southern China

never dried persimmons nor the Japanese in northern Japan never froze any. I presume they did, but not in large quantities.)

The Chinese had to rely on nature to do the freezing. With us it is much simpler. For freezing, the fruit should be washed and set out to dry. After drying they should be placed on a tray, no two fruits touching, and frozen in the freezer. They should then be stored in plastic bags in the freezer until used. If they are frozen in plastic bags or are in contact with each other, they freeze in a mass, which I find most undesirable.

A 1969 issue of Sunset magazine contained an article giving details on how to dry persimmons. Whoever wrote the piece was quite enthusiastic about the product. I have eaten dried Oriental persimmons and I don't share the author's enthusiasm. The Japanese sell and eat them as we would eat a confection. They frequently are dusted with confectioner's sugar.

The Sunset article suggests paring the fruit before drying, but I can't recall ever seeing a skinless dried persimmon in Japan. I'm not even sure that they wash them before drying them. They simply string them up by piercing them, top to bottom, with a long needle and hanging them under the eaves of their houses to dry. They string a dozen or two on a string. Presumably the number depends on the number of fruits available, the height of the eaves or the length or strength of the string. On occasion they knead or squash the drying fruit somewhat to break down cell walls and hasten the drying.

If I were to dry persimmons, I'd wash the firm-ripe fruit in a dilute calcium or sodium hypochlorite solution (a tablespoon of Clorox or Purex to a gallon of water), then press or roll the fruit over the sharp projections of a coarse stainless steel grater to make many minute skin punctures. The punctures will hasten the ripening of the fruit and allow it to dry quicker. Occasional kneading and additional heat will hasten the process.

American persimmons, having a higher sugar content than Oriental persimmons, are readily dried. In fact one may see an occasional tree with some dried fruit persistently hanging on the twigs during January and February. The fruit is dark when dried, approaching black, and by my taste standards, sweet but tasteless otherwise.

As an addendum to Dr. Griffith's comments, here is the way to can persimmon purée:

Wash fruit and boil in just enough water to prevent sticking. Press through a sieve or food mill. Add sugar to taste or leave unsweetened. Reheat to boiling and pour into hot Mason jars to 1/2 in. of top. Seal. Process in boiling water. Pints for 15 min.; quarts for 20 min. The purée is darkened by the processing.

PHEASANT

Draw pheasants as soon as possible after shooting and let them cool rapidly. Don't toss a lot of birds together in a pile, because this retards the loss of body heat.

If you dry-pluck pheasants, the job must be done while they are still warm. After they cool, it is tough going. Pull the feathers out between your fingers against the grain. Rub off the down with your thumb.

Pick out pin feathers. Singe off the hairs that remain.

If pheasants are cold before you're ready to pluck them, dip them quickly in water heated to about 150°. Then let them drain for several minutes before plucking and singeing.

Pheasants can also be skinned. Hunters generally frown on this because they think it results in loss of moisture and flavor, but the birds actually do not seem to be affected in any way. Cut off legs and wings at the second joint. Lay the bird breast down and cut the skin at the back of the neck. Pull it outward and downward, slipping it over wings and legs. Leave the skin in the tail piece.

Allow pheasants to age in the refrigerator for 2 or 3 days before wrapping in aluminum foil or in a polyethylene bag and freezing. Store for 9 mo.

PHYSALIS

Physalis is also called groundcherry and husk tomato. Hawaiians call it poha. The plant is an annual vegetable related to the tomato. It produces small, sweet, seedy, yellow fruits about the size of large cherries, and they are neatly enclosed in papery husks.

You can use the fruits, after husking, to make a conserve like that described under *Tomato.* You can also make the jam below. Since it comes from Kathryn J. Orr, foods and nutrition specialist in the Hawaii Agricultural Extension Service, we use the Hawaiian name.

POHA JAM

Makes 4 8-oz. glasses

3 lb. poha
1/4 cup water
1 cup sugar per cup cooked poha
1 tbs. lemon juice

Husk and wash fruit. Combine with water in a kettle and cook slowly, for 30 min. Stir frequently. Remove and let stand for 5 or 6 hr. Then measure pulp and juice and combine with an equal quantity of sugar. Cook slowly, stirring occasionally, for 1 hr. Add lemon juice and continue slow cooking until product passes a jam test. Immediately pour into hot, sterilized glasses and seal.

FREEZING

Husk and wash fruits. Crush or leave whole. Mix 4 cups fruit with 1 cup sugar. Package, seal, and freeze. Store for 10–12 mo.

PIGEON, BAND-TAILED

Hunting of this pigeon, found in our western mountains, is rather severely restricted today, but is not prohibited. Handle and preserve like doves.

PIKE

Several fishes are called pike. Pickerels are small pikes. In all cases the fish are found in fresh water and are long, lean, and voracious. To freeze, prepare the fish in the usual way; wrap in aluminum foil and put in the freezer. Store for 4–6 mo. Pike and pickerel are also salted.

PINCHERRIES

Also called wild red cherries, these northern trees bear small, acid, red fruits which make an excellent, colorful jelly.

JELLY

Makes 10 8-oz. glasses

3 1/2 lb. pincherries
3 cups water
6 1/2 cups sugar
1 bottle liquid pectin

Stem and pit cherries. Place in kettle with water. Bring to a boil and simmer for 15 min. Extract juice. Measure out 3 cups and mix with sugar. Follow standard procedure for making jelly with liquid pectin.

PINEAPPLES

DRY STORAGE

Ripe pineapples can be stored for up to 4 wk. in a dark, well-ventilated storage room at 40° to 45°.

FREEZING

Peel and core fruit and cut out eyes. Slice, dice, or crush the flesh. Pack in containers without sweetening or cover with 30 percent sugar sirup (made with pineapple juice rather than water if it's available). Store for 10–12 mo.

CANNING

Peel, core, and cut out eyes; then cut fruit into slices or cubes. Cook for 10 min. in 30 or 40 percent sugar sirup. Pack in jars and cover with hot sirup to 1/2 in. of top. Seal. Process in boiling water. Pints for 20 min.; quarts for 25 min.

JUICE

Peel pineapple and cut out eyes. Cut into small pieces, place them in a jelly bag, and squeeze out juice. The Smooth Cayenne variety does not need sweetening. With other varieties, add 1 cup sugar to 10 cups juice. Pour into glass freezer jars, seal and freeze. Store for 3–4 mo.

JAM

Without added pectin
Makes 7 8-oz. glasses

1 large pineapple
3 cups sugar
3 tbs. lemon juice
1 1/2 lemons

Leave leaves on pineapple so you can hold it. Peel and remove the eyes. Then holding the pineapple upright, strip off the flesh from top to bottom with a fork until you reach the core. Discard this.

Peel lemons and cut into narrow strips 1/2 in. long. Measure 6 cups shredded pineapple and juice into a kettle. Add sugar and lemon rind. Let stand overnight. Then add lemon juice and cook slowly till mixture meets a jam test—about 2 hr. Ladle into hot, sterilized glasses and seal.

JAM

With liquid pectin
Makes 11 8-oz. glasses

1 medium-large pineapple
6 tbs. lemon juice
6 1/2 cups sugar
1 bottle liquid pectin

Shred pineapple as above and combine 5 cups with the lemon juice and sugar in a kettle. Follow standard procedure for making jam with liquid pectin.

PICKLED PINEAPPLE

Makes 3 8-oz. glasses

1 medium-large pineapple
2 cups sugar
2 cups water
1 cup vinegar
1 2-in. stick cinnamon
3 whole cloves
pinch of salt

Peel pineapple and remove eyes. Cut crosswise into 1-in. slices, remove core, and cut slices into pieces about 1 in. wide. Combine with sugar and water and boil for 10 min. Then remove fruit and add other ingredients to the sirup and boil until it is thick. Add pineapple to sirup and boil for 5 min. more. Then ladle into hot, sterilized glasses and seal.

CANDIED PINEAPPLE

See Chapter 10. Peel pineapples, cut out eyes, and slice.

SHERBET

1 small pineapple
1 lemon
2 oranges
4 cups milk
1 1/2 cups sugar

Cut up and shred pineapple. Measure out 2 1/2 cups. Squeeze juice from lemon and oranges. Mix all ingredients together until sugar dissolves. Pour into pan and freeze. Stir once when partly frozen.

ICE CREAM

See *Milk.*

PINEAPPLE GUAVA

The pineapple guava is a tropical evergreen shrub that bears 2-in., gray-green fruits in the fall. They taste rather like a pineapple.

JELLY

Makes 9 8-oz. glasses

3 lb. pineapple guavas
3 cups water
1/2 cup lemon juice
6 cups sugar
1/2 bottle liquid pectin

Wash and cut up fruit. Simmer with water and lemon juice till soft. Then crush and bring to just below the boiling point. Extract juice. Combine 4 cups with sugar. Follow standard procedure for making jelly with liquid pectin.

PINE NUTS

Pine nuts are the seeds that drop out of pine cones. All pines have seeds, of course, but those large enough to be eaten as nuts are found mainly in the pinyon pines of the Southwest, in Italian stone pines, and in Korean pines. Few of these nuts actually grow very large. If they get as big as hazelnuts, it is rather unusual. But people who like them prize them. And the Indians of the Southwest not only have been depending on them to some extent for food for centuries but also today make money collecting and selling them.

Collect pine nuts from the ground under the trees. If damp, dry them in an airy place. Then store them in sacks, cans, what have you. They will keep for a long time. Nuts from the Colorado pinyon pine can even be stored for 3 yr.

PISTACHIO NUTS

Pistachio nuts come from an evergreen tree that grows in warm climates, particularly the inner valleys of California. When the husks rub off easily, the nuts are ripe. Knock them to the ground and remove the husks as soon as possible. Allow the nuts to dry for several days in a warm, airy place. Then store in sacks in a cool, dark place. Shelled nuts can be frozen or canned. See directions in Chapter 8.

PLOVER

Draw this small game bird as soon as possible after shooting. Then dry-pluck and cut off feet, outer wing bones and neck. Age for 24 hr. or more in the refrigerator. Then freeze and store for up to 9 mo.

PLUMS

FREEZING

Wash plums, slice in half, and remove pits from freestone fruits. Clingstone fruits can be frozen whole. Pack in containers and cover with 40 percent sugar sirup with ascorbic acid added. (Use 4 tsp. ascorbic acid per 4 cups sirup.) Freeze. Store for 10–12 mo.

CANNING

Raw pack. Wash fruit. Halve and pit freestone varieties. Leave clingstone varieties whole but prick skins with a fork. Pack in jars to 1/2 in. of top. Cover with boiling 40 percent sugar sirup to 1/2 in. of top. Seal. Process in boiling water. Pints for 20 min.; quarts for 25 min.

Hot pack. Prepare fruit as above and bring to a boil in 40 percent sugar sirup. Pack in jars and cover with boiling sirup. Leave 1/2 in. head space. Seal. Process in boiling water. Pints for 20 min.; quarts for 25 min.

PRUNES

Dry whole plums. Dip in boiling water for 1 or 2 min. to crack skins. Spread in single layers in trays and dry in oven for 18 to 24 hr. till wrinkled. Start at 130° and gradually increase to 165°. Store in tight containers.

JUICE

Wash, cut, and crush 2 lb. plums. Add 1 qt. water. Heat at 180° until fruit is soft. Then extract and strain juice. Add sugar to taste. Prepare juice by standard method.

JELLY

Without added pectin
Makes 4 8-oz. glasses.

3 1/2 lb. plums
1 1/2 cups water
3 cups sugar

Wash, cut plums in pieces, and crush. Combine with water, bring to a boil, and simmer till fruit is soft—about 10 min. Extract juice. Measure 4 cups into a kettle and mix in sugar. Follow standard procedure for making jelly without added pectin.

JELLY

With added pectin
Makes 9 8-oz. glasses

4 lb. plums
1 cup water
6 1/2 cups sugar
1/2 bottle liquid pectin

Wash and cut plums in pieces. Crush. Add water, bring to a boil, and simmer for 10 min. Extract juice. Measure 4 cups into a kettle and add sugar. Follow standard procedure for making jelly with liquid pectin.

JAM

Without added pectin
Makes 4 8-oz. glasses

2 lb. plums
2 cups water
3 cups sugar

Wash and pit fruit and chop in small pieces. Bring to a boil in water and simmer for 5 min. Mixture can then be put through a food mill or left as is. Mix in sugar. Follow standard procedure for making jam without added pectin.

JAM

With added pectin
Makes 10 8-oz. glasses

3 lb. plums
1/2 cup water
7 1/2 cups sugar
1/2 bottle liquid pectin

Wash and pit plums and chop in small pieces. Add water, bring to a boil, and simmer for 5 min. Measure 4 1/2 cups into a kettle and add sugar. Follow standard procedure for making jam with liquid pectin.

PRESERVES

Makes 6 pints

5 lb. plums
9 cups sugar
1 cup water

Wash sound, just-ripe fruits and pierce them with a fork. Put in a glass or similar container, cover with sugar, and add water. Cover and let stand overnight in a cool place. Then drain off sirup and boil it for 5 min. Add plums and cook until they are clear—only a few minutes. Pack in Mason jars, cover with sirup, and seal. Process in boiling water for 15 min.

CANDIED PLUMS

See Chapter 10. Prick skins with a sharp fork, cut fruits in half, and pit.

POKE

The Pennsylvania Dutch grow poke; we don't know anyone else who does. But you'll probably find it growing along road edges and in other wild areas. It is an interesting, tasty, asparaguslike vegetable.

To harvest, cut the shoots in the spring when they are 6 to 8 in. tall. Make the cuts just above the soil line. Take care not to cut any of the root, which is very poisonous.

Follow standard freezing procedure. Blanch for 8 min. Wrap stalks of the same length in bunches in aluminum foil and put in the freezer. Store for 10–12 mo. To serve, boil for 15–20 min.

Poke can also be canned like greens (which see). It should, however, be steamed or boiled for about 10 min. before being packed into jars.

POLLOCK

The pollock is also known as Boston blue. It's a salt-water denizen of mild and pleasant flavor. Wrap and freeze it according to general directions in Chapter 14. Store for 5–7 mo. Pollock is also frequently salted.

POMEGRANATES

The pomegranate is a tropical deciduous tree that produces large fruits, usually red, and always filled with seeds covered with a juicy pulp.

JELLY

Makes 10 8-oz. glasses

12 pomegranates
7 1/2 cups sugar
1 bottle liquid pectin

Wash fruits, cut in half, and press out juice with an orange juice squeezer. Strain through cheesecloth or a jelly bag. Measure 4 cups into a large kettle and add sugar. Follow standard procedure for making jelly with liquid pectin.

GRENADINE

As put up commerically, grenadine is made with pomegranates and assorted added fruit flavors. You can concoct the sirup to suit your taste. Or simply make grenadine out of pomegranates alone. To do this, extract and strain juice as above. Then follow directions for making sirup in Chapter 9.

POMPANO

This is the star of that fabulous—and not so hard to make—New Orleans dish called Pompano en papilotte. To make sure you have an ample supply of the fish on hand in anticipation of the time you want to make another paper-bag concoction, wrap them in aluminum foil and store away in your freezer. They will keep for 5–7 mo.

POPCORN

Leave the ears on the stalks until the kernels are dry. Then bring the ears into a dry place, remove the shucks, and store the ears until the kernels fall off under very little pressure. You can either remove and bag the kernels at this point or leave them on the cobs till you're ready for them.

PORGIES

Porgies are chunky salt-water fish. One species is called scup; another is the sheepshead. All freeze well in aluminum foil. Store for 6–9 mo.

PORK

Because pork is more perishable than other meats, it should be processed within 3 days after slaughter.

FREEZING

Select top-quality cuts. Wrap them tightly in aluminum foil and freeze for 4–6 mo.

CANNING

Follow standard procedure for canning meat in a pressure canner. Excess fat should be removed from meat before processing. If meat is cut up, process pints for 75 min.; quarts for 90 min.

SMOKING

Cure and smoke pork according to directions in Chapter 4. Allow 2 days per pound for dry-curing hams and shoulders; 4 days per pound for brine-curing. In the latter case, however, the minimum drying time is 28 days.

Bacon is both dry-cured and brine-cured for 1 1/2 days per pound.

Smokehouse temperature should not exceed 100° for all pork. A smoking time of 30 to 40 hr. is usually required to produce the desired flavor and amber to mahogany surface color.

After smoking, all pork should be wrapped to keep out insects. Dry-cured hams and shoulders can be stored for a year or more in a dark, dry, cool, well-ventilated place. Dry-cured bacon does not keep so well, however, and should be stored for only about 4 mo.

Whatever the cut, brine-cured pork should be stored at normal air temperatures for only a few weeks. It will keep longer in the freezer; even so it will go rancid fairly quickly, so you should not store it too long. The maximum for hams and shoulders is 3 mo.; for slabs of bacon, 6 to 8 wk. Sliced bacon and ham should be stored for only 4 wk.

All cured and smoked pork products should be cooked before serving. (This is not necessary, of course, with the precooked products sold in supermarkets.)

An important question you should answer before smoking hams is what sort of ham you like. If it is simply a good, juicy, tasty ham like the best sold at meat counters, you will probably be well advised to follow the schedule used on most farms: Cure and smoke your hams in the late fall or early winter so that they will not be attacked by insects during storage. (Another reason why farmers follow this schedule is that they don't have to refrigerate the hams during curing.)

If you prefer old country hams of the Smithfield type, however, you may want to follow a completely different schedule.

The thing that makes Smithfield hams different is that they are allowed to age before they are sold and eaten. Aging can be done at any temperature above about 40°, but it takes 8 to 12 mo., and by that time the ham may be pretty hard. Better results are attained when hams are aged for brief periods at fairly high temperatures—70° to 100°. This means that you should try to cure and smoke your hams during April and May and age them (let them hang) in a dark place at normal air temperatures during June, July, and August. Then store them in a cool place.

During this fast-aging process, hams need air circulation, so wrap them only in cloth. Air-tight wrappings often cause molding and spoilage. To keep insects away from the hams, clean and disinfect the aging room very carefully and screen all doors and windows with No. 30 mesh.

SAUSAGE

There are enough different sausage recipes to fill a whole book. We have space for only a few easy, representative types:

FRESH SAUSAGE

2 1/2 lb. fat pork
5 1/2 lb. lean pork
10 tsp. salt
8 tsp. ground sage
4 tsp. ground blackpepper
2 tsp. ground nutmeg
2 tsp. sugar

Use fresh pork trimmings. Mix the fat and lean together and put it through the coarse blade (approximately 1/2-in. holes) of a meat grinder. Then mix seasonings together and work them as evenly as possible into the meat. Grind the meat a second time using a blade with 1/4-in. holes (or 1/8-in. holes if you prefer a sausage with smooth texture). You can then stuff the sausage into casings available from a butcher or butcher supply house, but there is nothing to be gained by this. An easier procedure is to mix the sausage with a scant cup of cold water to make it less crumbly and more doughlike. Then form the sausage into patties and pack them into a large rigid plastic container. Separate each layer with aluminum foil. Freeze. Store for 5–6 wk. Cook thoroughly before serving.

Although it may not be so convenient, a better method of packaging—because there is less air in the package—is to stack the patties one atop another with aluminum foil in between, and then wrap the roll in foil.

If you want to keep the sausage for 2–3 mo., omit the salt from the recipe and pack the sausage in bulk in small rigid containers or aluminum foil. Then mix the salt into the sausage after defrosting, prior to cooking.

For even longer storage, sausage should be canned rather than frozen. Omit the sage from the recipe since it will make the sausage bitter. Form the meat into patties and cook them in a 325° oven until the meat at the center shows almost no red color. Drain off all fat and dry the patties on paper towels. Pack into clean, hot Mason jars to 1 in. of top and cover with a boiling broth to 1 in. of top. Seal. Process in a pressure canner at 10 lb. pressure. Pints for 75 min.; quarts for 90 min.

SMOKED SAUSAGE

Use the recipe above but increase the salt to 4 tbs. After grinding and mixing with water, stuff the sausage tightly into casings and tie the ends. Then place the sausage in your refrigerator for 24 hr. to cure. Smoke at 70° to 90° until the sausage is a dark mahogany color—about 1 or 2 days. Then let the sausage cool, wrap it in aluminum foil, and store in the freezer for up to 2 mo. Cook it thoroughly before serving.

CHORIZOS

- 3 1/2 lb. lean pork
- 3 1/2 lb. pork trimmings
- 3 lb. fat pork trimmings
- 7 1/3 tbs. salt
- 3/4 tsp. saltpeter
- 3 1/2 tsp. sugar
- 1 tsp. garlic powder
- 5 1/2 tsp. chile powder
- 4 tsp. red pepper
- 8 tsp. paprika
- 1 scant oz. white vinegar

Grind meat fine and mix thoroughly with vinegar and the cure (salt, saltpeter, and sugar). Pack 6 in. deep in containers and store in the refrigerator for 5 days. Add spices and mix thoroughly again. Then put in hog casings and hang to dry in a cool, well-ventilated place for 48 hr. or longer, till sausage looks quite dry. Then smoke for 10 to 12 hr. at 120°. To kill any trichina that may be in the pork, finish smoking for 30 min. at 165°. If you can't raise the smokehouse temperature this high, put the sausage in a 165° oven for the required time. Cool under cold water and store in a dark, cool, well-ventilated place. Because this is a dry sausage, storage under refrigeration is not necessary, though it won't hurt. Cooking before serving is unnecessary. But let the sausage cure for about 2 wk. after smoking before using it.

SCRAPPLE

7–8 lb. pork
1 1/2 cups buckwheat flour
5 1/2 cups corn meal
2 tbs. salt
2 tbs. ground black pepper
1 tsp. mace
1 tsp. ground nutmeg
1 tsp. sage

Use pork trimmings: head meat, feet, hearts, tongues. Cover meat with water and cook till done. Drain off and save broth. Pick over meat, removing bones, gristle, and excess fat. Grind fine. Measure out 5 lb., more or less, and combine it with twice as much broth. Bring to a boil.

Meanwhile, mix flour and corn meal with a little of the broth to make a smooth mixture, add to the meat and cook for 15 min. Then add seasonings and cook for 15 min. more, until thick. Stir constantly. Pour in small pans to chill and harden. Then remove and wrap in aluminum foil. Freeze. Store for 6 wk.

PICKLED PIGS' FEET

Clean pigs' feet thoroughly. Remove hair. Trim out glandular tissue between the toes. Then immerse the feet for 3 wk. in a brine made of 9 cups water, 1 lb. salt, 1/4 lb. sugar, and 1/4 oz. saltpeter. Weight the feet down. Store in the refrigerator. After curing, drain off brine, wash the feet in fresh water, and simmer in fresh water till meat is tender. Slow cooking is necessary to keep the skin from pulling away from the meat too much. When done, drain the feet and chill. Then pack in a crock under vinegar to which a few whole allspices have been added. The feet can be eaten within a few days. Store in the refrigerator for 3–4 mo.

LARD

Pork fat should be rendered not more than 7 days after the animal is slaughtered. Do not use intestinal fat; and since lard from leaf, or kidney, fat is better than that from other pork fat, it is advisable to render it separately.

Cut the fat into strips about 1 in. wide and remove the skin. Then cut the fat into small cubes or grind it. Place a small quantity in a large aluminum or stainless steel (but not copper) kettle and heat it slowly until it has started to melt. Add the remaining fat and continue cooking slowly. Stir frequently. Use a thermometer and don't let the fat get above 255° at the outside. A somewhat lower temperature is preferable.

As rendering proceeds, water in the fat will evaporate and the residual tissues (cracklings) will turn brown. These will float at first, then settle to the bottom. When most of them have settled, remove the fat from the range and allow it to settle and cool a bit. Then dip the top layer of fat into hot, dry, sterilized containers. Strain the bottom layer through two or three layers of cheesecloth into the containers. Fill the containers full. Chill them immediately in the refrigerator. Then cover them tightly and store them in a cool, dark place.

When removing lard for cooking purposes, leave as little as possible on the sides of the containers because it will eventually turn rancid and contaminate the rest of the lard.

The storage life of lard can be extended

somewhat if you mix 1 to 1 1/2 parts hydrogenated vegetable shortening with 25 parts lard just before the cracklings settle. Patented antioxidants may also be mixed with the lard according to the makers' directions.

If you wish, you may combine 1 part rendered beef fat with 9 parts rendered pork fat or 1 part rendered lamb fat with 20 parts rendered pork fat.

Lard that turns slightly rancid can be treated by heating about 10 lb. with 3 or 4 medium, sliced potatoes until the potatoes are a fairly dark brown. Then strain the lard into clean new containers.

POTATOES

DRY STORAGE

If you raise your own potatoes, dig up the tubers in the fall 2 wk. after the first frost has killed the tops. Do your digging on a day when the temperature is above 45°. Since the potatoes should not be exposed for long to sun or wind, collect them quickly, wash off dirt, let them dry, and put them in a basket or slatted crate in a dark, humid place at 50°. Potatoes held at this temperature will stay in good condition for 3–4 mo. For longer storage, hold the potatoes at 50° for 2 wk. after digging and then lower them to 38° to 40°. Complete darkness is essential to keep potatoes from turning green and poisonous.

If the storage temperature should drop below 38° for any length of time, the flesh will become too sweet. This can usually be corrected, however, by storing the potatoes at room temperatures for 2 wk. prior to cooking.

If it is impossible to store potatoes at adequately low temperatures, treat unbruised tubers with a sprout inhibitor at the time they are put into storage. The inhibitor, available from a well-stocked garden supply store, must be applied according to the manufacturer's directions.

FREEZING

To freeze potatoes, you must cook them first. The most common cooking method is French frying. Wash and pare the potatoes and let them stand in cold water for 30 min. Drain and dry on paper towels. Place a layer in a frying basket and fry in deep oil at 380° until they turn a light yellow—about 1 to 3 min. Drain on a paper towel and cool the potatoes as rapidly as possible in the breeze of a fan or in the refrigerator. Then pack in rigid containers and freeze. Store for 2–3 mo. To serve, spread in a single layer on a baking sheet and heat at 400° in the oven for 15–20 min.

Mashed potatoes stuffed in the shells or in small aluminum pans can also be frozen. Wash large potatoes and bake at 350° to 450° until soft. Cut off the top third. Scoop out flesh and put through a potato masher. Then gradually blend in hot milk—about 1/2 cup for 6 potatoes—and whip hard until fluffy. Add salt to taste. Cool completely in the refrigerator or by placing the pan in cold running water. Then stuff into bottom two-thirds of shells. Wrap in aluminum foil or place in rigid containers. Freeze at once. Store for 6–8 mo. To serve, heat in 400° oven for 30 min.

CANNING

Use potatoes which are 1 to 2 1/2 in. across. Wash and peel. Boil for 10 min. Pack in jars to 1/2 in. of top. Add 1/2 tsp. salt to pints; 1 tsp. to quarts. Cover with cooking water to 1/2 in. of top. Seal. Process in pressure canner at 10 lb. pressure. Pints for 30 min.; quarts for 40 min.

DRYING

Wash and peel potatoes and cut into small cubes or thin slices. Blanch in boiling water for 5 min. Then dry in oven at 140° until brittle. Store in tight containers in a dark, dry, cool place. The dried potatoes may be crushed into a powder.

POTATO CHIPS

Let potatoes cure for 2 wk. at 70°. Then peel, slice thin, and wash to remove starch. Cook till brown in vegetable oil. Dry thoroughly and sprinkle with salt. Package in tightly sealed plastic bags.

POT MARIGOLD

The pot marigold, or calendula, is an old-fashioned annual flower. The yellow petals of the blossoms are used fresh or dried in salads and soups. In the past they were used to make a medicinal tea. Collect the flower heads soon after they reach full size, and dry them in an airy, warm place. Then package in tight containers or polyethylene bags.

PRICKLY PEARS

Prickly pears are also called Indian figs and tuna. The pear-shaped, usually yellow, prickly fruits are borne in summer and fall on a large cactus belonging to the Opuntia genus. The plants are most commonly found in the Southwest but also grow and sometimes fruit in colder climates. These methods of preservation are recommended by Shirley L. Weik of the Arizona Agricultural Extension Service. We also owe much of our information about drying to Miss Weik.

JELLY

Makes 4 8-oz. glasses

2 lb. prickly pears
1 pkg. powdered pectin
3 tbs. lemon juice
3 1/2 cups sugar

Wear heavy gloves when handling pears. If you want to remove the bristles, rub them off with straw or leaves; but it is not necessary to do this. Simply wash fruits and put in a kettle with just enough water to cover. Boil until soft. Extract juice. Softened bristles will come off in the jelly bag. Let sediment in juice settle, and use only the clear juice.

Measure out 2 1/2 cups and add the powdered pectin. Bring to a fast boil, stirring constantly. Add lemon juice and sugar. Bring to a hard boil and boil for 3 min. Remove from range, skim, and pour into hot, sterilized glasses. Seal.

PRESERVES

Makes 4 8-oz. glasses

4 lb. prickly pears
1 1/2 cups sugar
5/8 cup water
2 1/2 tbs. lemon juice
1 orange slice 1/4 in. thick

Rub off bristles, wash fruits, peel, cut in half, and remove seeds. Combine with other ingredients and cook until fruit is transparent. Remove orange slice. Ladle into hot, sterilized glasses and seal.

PRICKLY PEAR PICKLES

Makes 4 half-pints

4 lb. prickly pears
2 cups sugar
2/3 cup cider vinegar
3 oz. red cinnamon candies

Rub off bristles, wash pears, peel, cut in half lengthwise, and remove seeds. Combine with other ingredients and cook until fruit is transparent. Ladle into hot, sterilized jars and seal. If you wish, you can cook 6 cloves in the sirup, but put them in a bag so they can be removed before bottling the pickles.

PUMPKIN

In Lyme every fall we have a pumpkin-growing contest which is becoming famous. The pumpkins can be entered by anyone in the area who has a garden, and the object is to see who can raise the largest. All the entries are sold for jack-o'-lanterns, and the proceeds go to maintain one of the ancient local cemeteries.

Unhappily, while the pumpkins are huge—the winner last year weighed 165 lb.—they are good only to look at. If you want good pumpkins for pies, stick to small ones. If you pick them before a killing frost, they can be stored in a dry, dark, airy place at 50° for 2–6 mo.

FREEZING

Use mature pumpkins without a stringy, coarse texture. Wash. Cut into pieces of fairly uniform size and scrape out seeds. Bake in a 350° oven until tender. Cool thoroughly in a pan in cold water. Then scoop pulp from rind and put it through a ricer or food mill. Pack immediately in rigid containers and freeze. Store for 10–12 mo.

If you want to prepare pumpkin pie filling before freezing, do so, using your favorite recipe. Then package and freeze.

CANNING

Wash pumpkin, cut into uniform pieces, and scrape out seeds. Bake in a 350° oven till tender. Remove pulp, and put through a ricer or food mill. Add a little boiling water to thin pulp to consistency needed for pie. Pack into jars to 3/4 in. of top. Seal. Process in pressure canner at 10 lb. pressure. Pints for 65 min.; quarts for 80 min.

QUAIL

Quail, or bobwhites, should be drawn as soon as possible after shooting. Allow them to cool rapidly. Then dry-pluck and trim off neck, lower legs, and outer wing bones. Age in refrigerator for at least 24 hr. Then package in aluminum foil or polyethylene bags, freeze, and store for 9 mo.

QUINCES

Quinces are shrubby deciduous trees that grow in temperate climates and produce large yellow or orange fruits resembling apples. They are not widely grown, and as a rule they are not well cared for; consequently the fruits are often riddled by worms and insects. But they make a jelly as delicious as the jelly made from sound fruits—which is to say that there is probably no jelly more delectable.

The quince tree is not related to the more familiar flowering quince, which also produces fruits that can be made into jelly. The latter are rarely used for this purpose, however.

JELLY

Without added pectin
Makes 4 8-oz. glasses

3 lb. quinces
4 1/2 cups water
1/4 cup lemon juice
3 cups sugar

Wash fruits and remove stem and blossom ends. Cut out bad spots. Grind up what is left (including cores), combine with water, and bring to a boil. Simmer for 15 min. Extract juice. Measure 3 3/4 cups into a kettle and add lemon juice and sugar. Follow standard procedure for making jelly without added pectin.

JELLY

With added pectin
Makes 10 8-oz. glasses

3 lb. quinces
4 1/2 cups water
1/4 cup lemon juice
7 1/2 cups sugar
1/2 bottle liquid pectin

Wash, trim, and grind fruit as above. Bring to a boil with the water and simmer for 15 min. Extract juice. Measure out 4 cups and combine with lemon juice and sugar. Follow standard procedure for making jelly with liquid pectin.

PRESERVES

Makes 4 8-oz. glasses

3 lb. quinces
3 cups sugar
2 qt. water

Wash, pare, quarter, and core fruit and discard bad parts. Boil sugar and water together for 5 min. Add fruit and cook until sirup is just below jelly stage. Stir often. Put fruit in hot, sterilized glasses and cover with sirup. Seal.

RABBITS

A word of caution about wild rabbits: a few of them have tularemia—rabbit fever—a serious disease which can be transmitted to humans through breaks in the skin. You can usually identify infected rabbits because they are sluggish or erratic. They may also have lesions under the skin or on the liver.

Most wild rabbits are healthy, however. Handle them in the same way as domesticated rabbits.

Remove entrails immediately after killing. Make an incision down the belly from vent to ribs. Be careful not to cut the intestinal casing. Clean out the body cavity thoroughly and wipe it with cleansing tissues. Cut off the head. Don't remove the skin until you are ready to process the rabbit because it keeps the flesh clean. To remove the skin, make a 2-in. cut through the skin across the middle of the back. Insert your fingers under the two flaps and pull firmly in opposite directions. Then cut off the feet with adhering skin. Wipe the carcass with a clean cloth dipped in vinegar and water to remove the hairs.

FREEZING

Cut rabbit into serving pieces and wrap them together in aluminum foil. Put in freezer. Store for 10–12 mo.

CANNING

Trim off as much fat as possible. Cut into convenient pieces. Make a boiling broth of scraps and bony pieces. Cook meat in this mixture, covered, until pink color is almost gone. Pack pieces loosely into jars to 1 in. of top. Add salt if desired: 1/2 tsp. for pints; 1 tsp. for quarts. Cover with boiling broth to 1 in. of top. Seal. Process in pressure canner at 10 lb. pressure. Pints for 65 min.; quarts for 75 min.

RACCOON

Raccoons are cleaned and beheaded soon after killing, but should then be allowed to hang for 24 to 48 hr. in a cool place before they are skinned. The meat is dark and coarse. Cut into serving portions; trim off excess fat; wrap in aluminum foil and freeze. Store for 6–8 mo.

RAIL

Rails are marsh birds. Draw them as soon as possible after shooting. Remove feathers by dry-plucking. Cut off lower legs, outer wing bones, and neck. Age in the refrigerator for at least 24 hr. Then freeze in aluminum foil. Store for 9 mo.

RASPBERRIES

FREEZING

Sort berries, wash, and drain thoroughly. Mix 4 to 5 lb. fruit with 1 lb. sugar. Do not stir too vigorously if you want more or less whole fruit. Pack in rigid containers and freeze. Store for 10–12 mo.

JUICE

Wash and crush fully ripe berries. Heat to 175°. Extract juice and strain. Add a little sugar only if you like a very sweet juice. Process by standard juice-making procedure.

SIRUP

See Chapter 9.

JAM

Makes 9 8-oz. glasses

2 qt. raspberries
6 1/2 cups sugar
1/2 bottle liquid pectin

Wash and crush berries. Sieve out some of the seeds if you wish. Measure 4 cups into a kettle and combine with sugar. Follow standard procedure for making jam with liquid pectin.

FROZEN JAM

Makes 6 8-oz. jars

1 qt. raspberries
4 cups sugar
2 tbs. lemon juice
1/2 bottle liquid pectin

Wash and crush berries. Measure 2 cups into a bowl and mix in sugar. Mix lemon juice with pectin in a separate bowl and add to fruit. Stir for 3 min. Pour into clean, cold jars and seal. Let stand at room temperature for 24 hr. Then put in freezer.

RED RASPBERRY JELLY

Makes 10 8-oz. glasses

3 qt. red raspberries
6 1/2 cups sugar
1 bottle liquid pectin

Crush berries, put in jelly bag, and extract juice. Measure 4 cups into a kettle and combine with sugar. Follow standard procedure for making jelly with liquid pectin.

BLACK RASPBERRY JELLY

Makes 7 8-oz. glasses

2 1/2 qt. black raspberries
1/4 cup lemon juice
5 cups sugar
1/2 bottle liquid pectin

Wash and crush berries. Heat them until juice flows; then simmer for 10 min. Extract juice. Measure 3 cups into a kettle. Add lemon juice and sugar. Follow standard procedure for making jelly with liquid pectin.

RASPBERRY ICE

2 cups sugar
2 cups water
3 cups raspberry juice

Boil sugar and water until sugar dissolves; cool thoroughly. Combine with raspberry juice, pour into a pan, and freeze in the refrigerator or freezer.

RED SNAPPER

Any time that a lot of dyed-in-the-wool seafood eaters get together, you are bound to hear words of high praise for this handsome fish from the Gulf of Mexico. Wrap and freeze it according to general directions in Chapter 14. Store for 6–9 mo.

RHUBARB

FREEZING

Wash, trim, cut into 1-in. pieces, and pack tightly in rigid containers. Seal and freeze. If you prefer a sweetened product, mix 1 lb. sugar with 4 to 5 lb. fruit.

If you generally eat rhubarb only as a stewed sauce, wash, trim, and cut stalks into 1-in. pieces. Place in a kettle with just enough water to prevent burning, cover, and cook gently till soft—about 20 min. Add sugar to taste and stir until dissolved. Let the rhubarb cool completely. Then pour into containers and freeze.

Store rhubarb for 10–12 mo.

CANNING

Wash and trim rhubarb and cut into 1-in. pieces. For each quart add 1 cup sugar, blend well, and let stand for 4 hr. Heat to boiling, stirring often, and boil for 30 sec. Pour into jars to 1/2 in. of top. Seal. Process in boiling water for 10 min.

JAM

Makes 8 8-oz. glasses

2 lb. rhubarb
3/4 cup water
5 1/2 cups sugar
1/2 bottle liquid pectin

Use bright red stalks; otherwise add a few drops of red food coloring. Chop into small pieces. Bring to a boil in kettle with the water and simmer until soft. Measure 3 cups into kettle and mix with sugar. Follow standard procedure for making jam with liquid pectin.

WINE

Cut washed rhubarb into 1/2-in. pieces, place in primary fermenter, and cover with water. Mix in 3 lb. sugar per gallon of the mixture and 1 package dried wine yeast. Then follow procedure outlined in Chapter 16. Add just enough additional sugar to make a mildly sweet juice, which will result in a dry, rose wine.

RICE

Rice is threshed and winnowed like other cereals. The hulled grains produced by the process are brown, not white, and have much higher nutritional value than the polished white grains. They also have a considerably shorter storage life.

The grains are usually boiled and eaten whole as a vegetable, but you can make coarse rice flour in a grist mill.

ROCAMBOLE

Rocambole is a perennial onion producing small, purple-skinned, garlic-flavored onions in the top of the plant. Harvest the bulbs in midsummer when the clusters begin to separate. Cut stems below the bulbs, tie them together, and hang them upside down in a dry, dark, cool place.

ROCK CORNISH GAME HENS

FREEZING

Pick birds when dry; then draw, wash, and cut off heads and feet. Chill thoroughly if freshly killed. Then wrap in aluminum foil or polyethylene bags and freeze. Store for 6–8 mo.

CANNING

Follow standard canning procedure for meat and poultry. If birds are small, it may be possible to pack them whole in wide-mouthed Mason jars. Process by the hot-pack method. Pints for 65 min.; quarts for 75 min. Otherwise, cut the birds into pieces and process by either the hot-pack or raw-pack method. Allow the same time for processing.

ROCKFISH

Several species of rockfish are caught along the Pacific Coast. The flesh varies from white to pink and has a surprising resemblance to crabmeat. Wrap the fish in aluminum foil and freeze. Store for 6–9 mo.

ROSE GERANIUM

This favorite, old-fashioned geranium has wonderfully fragrant leaves that can be dried, crumbled, and stored in bottles. Use them to make tea and to flavor biscuits.

ROSE GERANIUM JELLY

Make apple jelly according to directions under *Apples*. When almost done, add 3 or 4 rose geranium leaves. They will wilt in a few seconds and give off their flavor. Remove them at the last minute before packaging jelly. If you want a more colorful jelly, add a few drops of red food coloring.

ROSE HIPS

Rose hips are the red, berrylike fruits that develop on roses after the flowers fade. They are very rich in Vitamin C. The size of the fruits varies with the species of rose. Those from the dense, very thorny *Rosa rugosa* are among the largest.

Harvest the hips as soon as they turn bright red.

DRYING

Wash, cut open, and remove seeds. Spread the hips in a single layer on trays and heat in a 150° oven until completely dry. Put through a food grinder and then a sieve to make a powder for use in breads.

JELLY

Makes 7 8-oz. glasses

2 qt. rose hips
6 semitart apples
red and yellow food coloring
1/2 bottle liquid pectin

Wash and stem rose hips, place in a kettle with water to cover, and boil till soft. Wash and stem apples, cut into pieces, and boil till soft in another kettle. Extract juice from both fruits. Combine 2 cups rose hip juice, 3 cups apple juice, and the sugar in a large kettle. Add a few drops of the food colorings. Then make jelly by the standard procedure using liquid pectin.

SIRUP

Extract juice from rose hips as in making jelly. Then make into sirup according to directions in Chapter 9.

ROSELLE

Roselle is a tall, annual species of hibiscus, which is grown in our warmest climates for its red calyces. They resemble rose hips and have the flavor of cranberries.

JELLY

Makes 3 8-oz. glasses

2 lb. roselles
6 cups water
2 1/4 cups sugar

Wash and bring roselles to a boil in water. Cook till soft—about 7 min. Extract juice. Measure 3 cups into a kettle with the sugar. Follow standard procedure for making jelly without added pectin.

SAUCE

Remove stems and seeds from calyces. Measure calyces and mix with an equal amount of water. Cook slowly till tender. Put through a food mill. Add 1 cup sugar to 6 cups pulp. Bring to a boil and pour into hot Mason jars to 1/4 in. of top. Seal. Process in boiling water. Pints for 8 min.; quarts for 12 min.

JUICE

Extract juice as for jelly. Chill and mix 4 cups with 1 cup sugar. Pour into rigid containers, seal, and freeze. Store for 3–4 mo.

ROSEMARY

Rosemary is a large, evergreen perennial herb that grows outdoors the year round only in warm climates; elsewhere it should be grown in pots and brought indoors in winter. The small, dark green leaves are used fresh or dried to season a variety of foods; and the dried leaves also make a tea. Harvest the leaves at any time—but preferably before the blue flowers appear. Dry them in an airy place and pack in tight containers.

RUTABAGAS

Rutabagas are close relatives of turnips and very popular with northerners. The big, rough roots are purple on top, amber below. The flesh is yellow and sweet.

DRY STORAGE

Harvest after the first frost but before the ground and roots freeze. Keep in moist sand in your dry storage area at 33° to 40°. For the longest possible life, clean the roots after harvesting and brush them with melted paraffin before storing.

FREEZING

Use young, medium-size roots. Remove tops, wash, and peel. Cut into pieces and boil until tender. Drain, mash, and cool in a pan set in cold water. Pack into rigid containers, seal, and freeze. To serve, heat in a double boiler.

Another way of preparing rutabagas is to dice the roots in 1/2-in. pieces. Follow standard freezing procedure. Blanch for 2 min. To serve, boil for 12–15 min.

Rutabagas frozen by either method can by kept for 10–12 mo.

RYE

Rye has more weather hardiness than any other cereal grain and is, accordingly, grown more in cold climates than in warm. Thresh, winnow, and store the grain in sacks in a dry place. Grind into flour as needed or in larger quantities in anticipation of need. Store flour in tight containers.

SABLEFISH

Also called black cod, sablefish is caught in the North Atlantic. It averages between 5 and 10 lb., has fine, white, flaky flesh. Freeze in aluminum foil and store for 6–9 mo.; or cure and give it a hot smoke.

SAFFLOWER

Safflower, or saffron, is a prickly annual plant with thick clusters of yellow flowers. Their petals are used as a food coloring substitute for the real saffron. Cut the flowers on a dry day when they are fully open, place them on screen cloth, and dry them in an airy place out of the sun. Then put the petals in a tight container.

Safflower is grown mainly for its hard, dry seeds, which are crushed to yield an oil. However, we have not been able to discover any satisfactory method of crushing them at home.

SAGE

Sage is a perennial herb with aromatic, gray-green leaves used in dressings, sausages, stews, cheese, and many other things. Harvest the most succulent young leaves before flowering starts; dry them in an airy, warm place, and pack into tight containers.

As sage plants age they become woody. Leaves taken from these plants are less aromatic than those from young plants.

SALAL

Salal is a low evergreen shrub growing on the Pacific Coast. It is loaded with hairy-husked, purple-black berries about the size of blueberries. To make a very good jelly, wash, crush, and cook them gently for a few minutes. Then extract juice. Measure and combine with an equal quantity of tart apple juice. Bring to a boil and add 1 cup sugar for each cup juice. Follow standard procedure for making jelly without added pectin. One cup of the combined fruit juices yields a little less than 1 glass jelly.

SALMON

FREEZING

New England tradition has it that you should eat poached salmon and fresh peas on the Fourth of July—and what a glorious way that is to celebrate the Fourth. To put up salmon in anticipation of the date, leave it whole or cut it crosswise in half. Wrap in aluminum foil and freeze.

Salmon steaks are wrapped and frozen in the same way. If you package several together, separate them with sheets of wax paper or aluminum foil.

However it is packaged, salmon may be stored for 6–9 mo.

CANNING

After preparing in the usual way, cut fish into pint-jar lengths. Don't remove backbone. Immerse in a brine made in the proportions of 2 qt. water and a scant 1/2 cup salt, for 1 hr. Then drain thoroughly and pack solidly into jars, but without crushing. Fill to 1/2 in. of top. Seal. Process in pressure canner at 10 lb. pressure for 110 min.

SMOKING

Follow standard curing and smoking procedure or take your leaf from the Northwest Indians and simply smoke salmon over a fire until it is dry. Hot smoking in a smokehouse is preferred to cold smoking.

Smoked salmon should be double-wrapped and frozen. It may also be canned. See directions in Chapter 4.

SALTING

See Chapter 5.

PICKLED SALMON

10 lb. boned salmon
1 qt. water
1 cup sliced onions
1/2 cup olive oil
1 qt. white vinegar
10 bay leaves
1 tbs. white pepper
1/2 tbs. black peppercorns
1 tbs. mustard seed
1/2 tbs. cloves

Cut salmon into small serving portions, wash, drain, and dredge with salt. Let stand for 30 min., then rinse and simmer in water until done. Drain and place in a sterilized crock or Mason jars.

Meanwhile cook onions in oil until yellow. Add other ingredients and simmer for 45 min. Cool and pour over fish. Hold fish down, if necessary, with a piece of crushed plastic film. Seal tightly. Store in the refrigerator.

SALMONBERRIES

A close relative of the raspberry, the salmonberry grows mainly in the western wilds. The berries are not very distinguished when eaten raw, but make a fine preserve. Pick all you can on your next hike; wash and measure; and combine with an equal amount of sugar. Cook slowly for 15 min.; then remove from the burner and spoon the berries into another container. Boil the remaining juice for another 15 min. Add the berries, lower the heat, and stir slowly till the berries are heated through. Then ladle into hot, sterilized glasses and seal. One quart berries yields approximately 2 glasses preserves.

SALSIFY

Salsify is also known as oyster plant and vegetable oyster because the creamy-white flesh of its 8-in., carrotlike roots taste rather like oysters. The easiest way to store the roots is to leave them in the ground and dig as needed. Or dig up the roots in the fall and store them in boxes of damp sand in a cold root cellar.

FREEZING

Remove tops. Wash and scrape roots and cut into 1-in. pieces. If you are not going to blanch the vegetable at once, put the pieces in a mild solution of lemon juice and water to prevent discoloring. Follow standard freezing procedure. Blanch for 3 min. Store for 10–12 mo. To serve, cook for 12–15 min.

SANDCHERRIES

The sandcherry is a northern shrub producing astringent purple-black fruits up to 1 in. across.

JELLY

Without added pectin

Wash, halve, and pit cherries. Weigh and add half this weight in water. Cook till fruit is soft. Extract juice. Mix 1 cup tart apple juice with each 2 cups cherry juice. Add 3/4 cup sugar for each cup mixed juice. Follow standard procedure for making jelly without added pectin. One cup combined juices yields about 3/4 glass jelly.

JELLY

With added pectin
Makes 10 8-oz. glasses

3 1/2 lb. sandcherries
3 cups water
6 1/2 cups sugar
1 bottle liquid pectin

Wash, halve, and pit fruit. Combine with water and bring to a boil; then simmer for 15 min. Extract juice. Measure out 3 cups, add sugar, and follow standard procedure for making jelly with liquid pectin.

SANDCHERRY-PLUMS

Sandcherry-plums, also called cherry-plums, are crosses between sandcherries and plums. Grown mainly in the Northern

Plains States, they vary from about 1/2 to 1 1/4 in. across, have purple, red or green skins, and yellow to purple flesh.

The fruits of the best, most commonly cultivated varieties are more like plums than cherries, and are preserved like plums.

SAND DABS

Sand dabs are flat fish closely related to flounders and resembling them. Freeze them whole or filleted in aluminum foil or polyethylene bags. Store for 6–9 mo.

SARDINES

Sardines, or pilchards, are small Pacific Coast members of the herring family. They are very good when pan-fried like smelts, and they can be packaged and frozen like smelts. Store for 5–7 mo.

Sardines are more often canned, however. Clean and behead them. Cut into short lengths as necessary, and pack cut pieces or whole fish vertically in pint Mason jars to 1/2 in. of top. Pack tightly, but do not crush. Cover with olive oil. Seal jars and process them in a pressure canner at 10 lb. pressure for 100 min.

SASKATOONS

Saskatoons, or serviceberries, are 10-ft. shrubs grown mainly in the Northern Plains States. They have purple-black fruits with a conspicuous bloom. They are used in desserts and pies and to make jam.

FREEZING

Wash and dry the berries completely. Pack dry in rigid containers, or crush them slightly and add 1 cup sugar to 6 cups fruit. Freeze. Store for 10–12 mo.

CANNING

Wash and boil for 5 min. in 30 percent sugar sirup. Pack into jars to 1/2 in. of top and cover with sirup to 1/2 in. of top. Seal. Process in boiling water bath for 25 min.

JAM

Makes 4 8-oz. glasses

3 lb. saskatoons
2 lemons
2 oranges
3 cups sugar

Wash and chop berries coarsely. Measure 4 cups into a kettle and just cover with water. Boil gently till fruit is tender. Add the juice of the lemons and the pulp of the oranges cut in tiny pieces. Then add 1 tsp. grated orange rind and sugar. Boil slowly for 20 min. Ladle into hot, sterilized jars and seal.

SCALLOPS

The scallop gets its name from the scalloped edge of its shell. In the United States we eat only the cylindrical muscle, called the eye, which propels the scallop through the water, but there is no reason why you shouldn't eat all the meat as Europeans do.

Most people are familiar only with the large sea scallop which is served deep-fried in restaurants, but the much smaller bay scallop (the eye is only about as large as a fingernail) is both sweeter and more tender.

To freeze scallops, wash them well. Open with a knife and cut out the muscle. Wash in 2 qt. cold water to which 1 tbs. salt has been added. Drain. Pack into rigid plastic containers or glass jars. Freeze. Store for 6–9 mo. To serve, thaw and cook until tender.

SEAGRAPES

The seagrape is a handsome tropical evergreen tree with big, leathery leaves and bunches of small, purple, grapelike fruits.

JELLY

Makes 9 8-oz. glasses

3 lb. seagrapes
1/2 cup water
7 cups sugar
1/2 bottle liquid pectin

Use half-ripe fruit. Wash, stem, and crush. Bring to a boil with the water and simmer for 10 min. Extract juice. Measure 4 cups into a kettle with the sugar. Follow standard procedure for making jelly with liquid pectin.

If the jelly is too mild, make it the next time with 3 1/2 cups juice and 1/2 cup lemon juice.

SESAME

Sesame, or bene, seeds are produced by a fairly tall annual plant growing in warm climates. The seeds are enclosed in capsules which grow in thick clusters around the stalks. Harvest the capsules as soon as the first begin to burst by cutting off the plants near the ground. Tie the plants in bunches and hang them upside down in a warm, dry place over cheesecloth to catch the falling seeds. Package seeds in bottles.

SHAD

Along about the end of April, after the shad have started their annual run up the Connecticut River, we take a ride down to see our favorite fish lady, who is named Jerry, and ask her how the run is going and how much she and her fisherman husband are charging for shad roes and fillets in quantity. Then we get on the long-distance phone to the Houstons, Richardsons, Dommerichs, Eggerts, and anyone else who is crazy about shad, and ask them what they want to order this year for freezing.

Later, when the run is at its height, everyone drives to Lyme for lunch. Then home we all go, each with a dozen or two pairs of roe, a somewhat smaller number of fillets, and perhaps a couple of whole buck shad. And we live happily for another year.

FREEZING

We've always had good luck just wrapping roes, fillets, and whole fish in freezer paper and overwrapping in polyethylene bags or aluminum foil. Everything keeps in the freezer in good shape for 9 mo. If you glaze the roes and fillets, however, they will keep a month or two longer.

SMOKING

Follow standard procedure for curing and cold-smoking fish. Store in the freezer, or can.

SALTING

See Chapter 5.

SMOKED ROE

Do not break the membrane surrounding the eggs. Wash the roe carefully in cold water, dredge it with salt, and let it cure in the refrigerator for 6 to 8 hr. Then wash in cold water and dip for 2 min. in boiling water. Pat dry with paper towels and place the roe on oiled wire mesh in the smokehouse. Expose to dense smoke for 6 to 8 hr. at 80°. Then cool completely; wrap in foil or plastic film and freeze. Store for 3–4 mo.

DRIED ROE

Wash the roe carefully but thoroughly in cold water. Don't break the membrane. Then drain it for 30 min., dredge with salt, and let it cure for 12 hr. in the refrigerator. Then brush off excess salt and place the roe in direct sunlight to dry. Turn every hour.

When night comes, bring the roe indoors to a cool place and place it under a board and weight to compress it slightly. Dry again the next day in the sun, but without turning, and weight it down once more that night.

Continue drying outdoors every day until the roe feels hard when you pinch it and is yellow to red-brown in color. This takes about a week. Bring the roe indoors every night and whenever the weather turns threatening during the day.

Brush melted paraffin or beeswax on the dried roe; let it cool briefly; then wrap in wax paper and store in a box in a cool, dry place. To serve the roe, slice it thin like sausage and eat it without further preparation.

SHADDOCKS

The shaddock, which is also called pummelo, is a large citrus tree closely related to the grapefruit. Its big fruits range from round to pear-shaped. They are rated inferior to grapefruits in this country, but they can be put up in the same ways.

SHALLOTS

Shallots are a type of onion which produces small, brown-skinned bulbs. Dig these up when the tops are mostly withered. Dry them in a cool, airy place. Store in a dry, dark, cold place.

SHARK

A number of members of this villainous family make surprisingly good provender when filleted or cut into steaks.

Since shark meat spoils very rapidly, it must be cleaned and processed very soon after catching. Remove the dark flesh and preserve only the light. One method of preservation is by freezing. Wrap the meat in aluminum foil, place in the freezer, and store for only 4–6 mo. because the meat is very fatty.

You can also dry-salt shark meat. The process requires about 10 days in salt and 10 days of drying. The meat is stored in wax paper in a cool, dark place. Or it may be smoked at 80° for 10 hr. before it is stored.

SHRIMP

FREEZING

Uncooked. Break off the heads, wash briefly in cold water, drain, and freeze in the shells. The alternative is to remove heads, peel off shells, and scrape out black veins. Then freeze. Store for 6–9 mo. To serve, boil for 12–15 min.

Cooked. Boil shrimp in salted water for 10 min., then remove shells and veins. Cool quickly in the refrigerator and pack in polyethylene bags. Freeze. Store for 3 mo. To serve, bring to a boil and let stand off the heat for 1 min.

CANNING

Remove heads, shells, and veins. Wash in fresh water, then immerse in a cold brine of 1 qt. water and 1/2 cup salt for 30 min. Drain and pat dry. Place in a deep-fat frying basket no more than to half its depth. Fill a large kettle with brine made in the previous proportions, bring to a boil, and lower the shrimp into it. Bring to a boil again and cook for 6 min. Remove shrimp and spread on wire mesh trays or open racks to dry and cool thoroughly under an electric fan. Then weigh shrimps in small quantities and pack 6 oz. into half-pint jars; 12 oz. into pint jars. Cover with a boiling brine made in the proportions of 1 qt. water and 1 1/2 tbs. salt. Leave 1/2-in. head space. Seal. Process in pressure canner at 10 lb. pressure. Half pints for 25 min.; pints for 35 min.

SKATE

The ugly skate is prized in Europe as a food, but it is not often eaten in the United States. You might like it, however—especially the wings of the little skate. Remove the wings and cut into manageable pieces. Wrap them in aluminum foil and freeze. Store for 6–9 mo.

SMELTS

Smelts are among the smallest of all fish we have included in this book, but if they are not the most delicious, they are very close to it. The white flesh has a sweetness and flavor all its own. And to make eating even better, you don't have to worry about the bones. Except for the backbone in an occasional very large fish, they go down about as easily as the flesh.

To clean smelts, cut off the heads, make a slit down the belly, and push out the entrails with your thumbnail. Wash well. Wrap in aluminum foil about a half pound of fish for each member of the family. Freeze. Store for 5–7 mo.

SNIPE

Snipes are long-billed shore birds. Draw them as soon as possible after shooting. Dry-pluck. Cut off feet, outer wing bones and neck. Age for 24 hr. or longer in the refrigerator. Then freeze and store for up to 9 mo.

SNOOK

Snook are salt-water fishes that often venture into fresh water. You'll find them only in our warmest waters. They are fun to catch and excellent for eating. Wrap and freeze in the usual way. Store for 6–9 mo.

SOLE

The sole is the European version of our flounder (there are a few soles in American waters, however), and generally rated a bit better to eat (though we're not convinced of this). Freeze whole fish or fillets in aluminum foil. Store for 6–9 mo.

SORGHUM

Sorghum is a widely cultivated member of the grass family and used for many things. Grain sorghums are fed to livestock. Sweet sorghums are made into an excellent sirup or molasses.

Harvest the stalks of the sweet sorghum when the seed is in the late-milk to soft-dough stage (test the seed with your thumbnail as you would test kernels of sweet corn). Cut the stalks 6 in. above the ground. Remove the leaves, seed heads, and two of the stalk's upper joints. Then press the juice from the stalks by running them through iron rollers (see *Sugar Cane*) or by beating them with the side of an ax.

Strain the juice and let it settle for a couple of hours; then pour off all but the bottom 1 or 2 in. and bring to a boil. Cook down to a sirup. Skim frequently, especially before the juice comes to a boil; 6 to 7 gal. juice yield 1 gal. sirup.

Cool the sirup as quickly as possible to about 140° by placing the kettle in cold water. Then let the sirup stand overnight. In the morning, pour off and strain the clear juice above the sediment; heat it gently and pour into hot, sterilized bottles or Mason jars. Place these, uncovered, in deep water and heat to 190° for 5 min. Then seal tight. Store in a dark place at room temperature.

SORREL

Sorrel is a perennial that is as likely to be found in weed patches as in gardens. The light green leaves have an acid taste which does interesting things for some sauces. But the best use we know of for sorrel is in soup, and that happens to be the only way we know how to preserve it.

SORREL SOUP

1/2 lb. sorrel
1/2 stick margarine
6 cups chicken consommé
4 egg yolks
1 cup heavy cream

Use young leaves. Pick over and cut out thick stems. Cut leaves into thin shreds. Brown lightly in margarine. Bring consommé to a boil, add sorrel and melted margarine, and simmer 20 to 25 min. Let cool thoroughly. Pour into rigid containers and freeze. Store for 6–9 mo.

When you're ready to serve the soup, if it is to be served cold, remove frozen packages and let them thaw completely. Beat egg yolks lightly and stir into the soup along with the cream. Serve. If you prefer hot soup, heat the frozen stock in the top of a double boiler; stir in beaten yolks and cream, and bring up to serving temperature. Then serve.

SOUR ORANGES

The superb marmalade that the English make is concocted with the sour, or Seville, orange.

MARMALADE

Makes 4 8-oz. glasses

6 sour oranges
4 cups water
3 cups sugar

Wash oranges and peel 2 of the oranges thinly, leaving as much of the white rag on the fruits as you can. Cover the peel with water and boil to remove the bitter oils. Drain. Quarter and seed the pulp of the 6 oranges, and put through a food chopper. Chop the cooked peel. Combine pulp and cooked peel with water and boil for 20 min. With a sieve, dip out roughly half of the pulp and peel and discard it. Measure out 3 cups remaining pulpy juice, bring to a boil, and add sugar. Boil to jam stage. Pour into hot, sterilized glasses and seal.

JELLY

Makes 3 8-oz. glasses

4 sour oranges
4 cups water
3 cups sugar

Wash and peel oranges, leaving as much of the white rag as possible. Discard peel. Quarter fruit, remove seeds, combine with water, and boil till fruit falls apart. Extract juice and strain a second time. Measure 3 cups into a large kettle, add sugar, and cook rapidly to jelly stage. Ladle into hot, sterilized glasses and seal.

SOURSOP

Despite its name, the soursop is considered to be a delicacy in the tropics where it grows. The fruits are large and long, dark green outside and creamy white inside. To freeze soft, ripe fruits for later use in beverages, sherbets, and ice cream, peel them and cut lengthwise through the center. Separate the numerous seeds from the flesh and throw them away. Purée the remaining pulp and mix 6 cups with 1 cup sugar. Package and freeze. Store for 10–12 mo.

SOYBEANS

Soybeans resemble lima beans and are cooked and served in the same way. The edible beans are not the same, however, as those grown on a vast scale for meal and oil. Bansei and Kanrich are two good varieties to use.

FREEZING

Pick beans when they are bright green, fat, and tender. Wash. Heat pods in boiling water for 5 min. Chill. Squeeze out seeds and pack them into rigid containers or polyethylene bags. Freeze. Store for 10–12 mo. To serve, cook for 12–14 min.

DRYING

You can harvest pods and shell them after they have dried thoroughly on the vine. But a better procedure is to pick young, green beans. Wash and heat pods in boiling water for 7 min. Squeeze out seeds. Spread in 1/2-in. layers in trays. Dry for 6 to 10 hr. in the oven, stirring constantly. Start oven at 115° and gradually increase to 140°. Store dried beans in tight containers in a cool, dry, dark place.

SPINACH

Freeze or can like other greens (which see).

SQUABS

Baby pigeons used to be considered a great delicacy and were served mainly at parties, but they have lost favor since Rock Cornish game hens came on the scene. This does not, however, demote squab from the delicacy category.

Pluck squab dry. Cool thoroughly. It is then best to wrap them tightly in aluminum foil or polyethylene bags and freeze for 6–8 mo. Whole squabs may, however, be canned by the hot-pack method in wide-mouthed Mason jars. Follow standard procedure for meat and poultry. Process in pressure canner at 10 lb. pressure. Pints for 65 min.; quarts for 75 min.

SQUASH, SUMMER

One of our biggest problems every summer is to find ways to use all the summer squash the garden produces. We eat a lot, and process a lot, and give a lot away—and still we have more. Of course, we could cut down on the number of plants we grow, and over the years we have done just that. But we've reached a point below which we can't make ourselves go. New uses must be found.

FREEZING

Use fruits with tender rinds that can be cut with a fingernail. Cut in 1/2-in. slices. Follow standard freezing procedure. Blanch for 3 1/2 min. Store for 10–12 mo. To serve, cook for 10–12 min.

CANNING

Cut young squashes into uniform pieces. Place in a kettle, just cover with water, and bring to a boil. Pack squash loosely in jars to 1/2 in. of top. Add 1/2 tsp. salt to pints; 1 tsp. to quarts. Cover with boiling cooking liquid to 1/2 in. of top. Seal. Process in pressure canner at 10 lb. pressure. Pints for 30 min.; quarts for 40 min.

RATATOUILLE

This is as good a way to make use of excess zucchini as it is to get rid of excess eggplants and tomatoes. Follow recipe under *Eggplant.* If you wish, you can omit the eggplant and use 2 zucchini.

BREAD AND BUTTER PICKLE

Makes 6–7 pints

1 qt. vinegar
2 cups sugar
3 tbs. salt
2 tsp. celery seed
2 tsp. turmeric
1 tsp. ground mustard
4 qt. sliced zucchini (about 6 slender 10-in. zucchini)
1 qt. sliced onions (about 1 lb.)

In a large kettle boil together vinegar, sugar, salt, and spices. Add vegetables and let stand for 1 hr., stirring occasionally. Bring to a boil again and cook for 3 min. Pack into hot, sterilized jars and seal.

SQUASH, WINTER

Winter squashes differ from summer squashes in that they are generally borne on vines which run across the ground and have hard rinds. They also mature more slowly and are not ready for harvesting until late summer, as a rule.

DRY STORAGE

If fruits are sound, they will keep for most of the winter in a well-ventilated, somewhat humid, dark storage space at a temperature of 50°. If the temperature drops below 50° for short periods, it does no great harm. However, prolonged exposure to below-50° temperatures or brief exposure to below-freezing temperatures will ruin the fruits. Temperatures between 55° and 60° are also undesirable because they cause shrinkage of the fruits, but they are less damaging than low temperatures.

FREEZING AND CANNING

Handle like pumpkins.

SQUID

The squid, or cuttlefish, is a sea-dwelling mollusk related to the octopus. It isn't pretty, but it is rich and sweet to eat. Clean and cut them into cubes about 2 in. across. Put in polyethylene bags or rigid plastic containers and freeze. Store for 6–9 mo.

SQUIRREL

Squirrel is a tender, delicately flavored meat with very little gamey taste. Dress and process like rabbit.

STRAWBERRIES

We were gardeners for many years before we got around to raising our own strawberries. And now that we have discovered what a good thing we were missing, we urge you to raise strawberries, too. They're easy. Twenty-five plants take up only 50 to 100 sq. ft., depending on how you train them. And the fruit they yield is not only beautiful and delicious but also plentiful. Each plant should give you a pint of berries.

All varieties are good for making jam, preserves, and the like. But for freezing use Catskill, Dunlap, Earlidawn, Fletcher, Northwest, Pocahontas, Redglow, Sparkle, or Surecrop.

FREEZING

Pick over and hull fruit and wash in very cold water. If packing whole berries, cover them with cold 40 percent sugar sirup or mix 4 to 5 lb. fruit with 1 lb. sugar. If you cut berries, mix with sugar at the same rate. Package in rigid containers and freeze. Store for 10–12 mo.

JUICE

Wash, hull, and crush berries. Heat to 175°. Squeeze through a jelly bag and then strain through cheesecloth. Sweetening is unnecessary, but if you prefer to use it, dissolve 1 cup sugar in 9 cups juice. Then follow standard juice-making procedure.

SIRUP

See Chapter 9.

ICE CREAM

See *Milk.*

JAM

Without added pectin
Makes 4 8-oz. glasses

2 qt. strawberries
4 cups sugar

Sort, hull, wash, and crush berries. Measure 4 cups into a kettle and combine with sugar. Follow standard procedure for making jam without added pectin.

JAM

With added pectin
Makes 9 8-oz. glasses

2 qt. strawberries
7 cups sugar
1/2 bottle liquid pectin

Sort, hull, wash, and crush berries. Combine 4 cups in a kettle with the sugar. Follow standard procedure for making jam with liquid pectin.

If strawberries lack tartness, use 3 3/4 cups crushed fruit and 1/4 cup lemon juice.

FROZEN JAM

Makes 5 8-oz. jars

1 qt. strawberries
4 cups sugar
2 tbs. lemon juice
1/2 bottle liquid pectin

Sort, hull, wash, and crush berries. Measure out 1 3/4 cups and mix with sugar. Mix lemon juice and pectin in a small bowl and add to fruit. Stir for 3 min. Pour into clean jars and seal. Let stand at room temperature for 24 hr. Then freeze.

JELLY

Makes 10 8-oz. glasses

3 qt. strawberries
1/4 cup lemon juice
7 1/2 cups sugar
1 bottle liquid pectin

Sort, hull, wash, and crush berries thoroughly. Place in a jelly bag and extract juice. Measure 3 3/4 cups into a large kettle and add lemon juice and sugar. Follow standard procedure for making jelly with liquid pectin.

PRESERVES

Makes 4 half-pints

1 1/2 qt. strawberries
5 cups sugar
1/3 cup lemon juice

Use firm berries. Sort, wash, and hull. Combine fruit and sugar in alternate layers and let stand for 4 hr. Then bring slowly to a boil, stirring occasionally, until sugar dissolves. Add lemon juice. Boil for 10 min. Do not stir kettle but shake and skim. Pour mixture into a shallow pan and let stand for 12 to 24 hr. in a cool place. Shake pan occasionally, especially before sirup cools, in order to plump up berries and distribute them through the sirup. Then bring berries to a rolling boil and pack in hot, sterilized jars. Seal and store in a dark place to keep berries from fading.

CONSERVE

Makes 10 8-oz. glasses

1 1/2 qt. strawberries
1 pineapple
1 cup seedless raisins
1 orange
sugar
2 tbs. lemon juice

Wash, hull, and crush strawberries. Peel and cut eyes out of pineapple; cut into pieces, chop fine and measure out 2 cups. Grind raisins. Peel and grind up orange. Measure the fruits into a large kettle with an equal amount of sugar, and add lemon juice. Cook till thick, stirring often. Ladle into hot, sterilized glasses and seal.

WINE

Mix 4 qt. hulled, sliced strawberries with 1 lb. sugar in an open fermenter. Allow it to ferment violently for about 8 days, till it subsides. Stir and poke the cake down every day. Then strain juice into a gallon jug and add 3 lb. sugar and one-fifth package of dried wine yeast. Fill jug to top with water. Cover with cheesecloth and let the juice ferment until it subsides considerably. Then close the jug with a fermentation lock. When bubbling stops, clarify the wine as necessary and then bottle it. Don't serve it until it is 6 mo. old.

STRAWBERRY GUAVA

A relative of the common guava, the strawberry guava is a tropical shrub with small red fruits. The plant is more commonly grown for ornament or as a hedge than for its fruit; but the latter is good nonetheless.

JELLY

Makes 9 8-oz. glasses

3 lb. strawberry guavas
3 1/2 cups water
1/4 cup lemon juice
6 cups sugar
1/2 bottle liquid pectin

Wash and cut up fruit in small pieces. Combine with water and lemon juice, and crush. Bring to just below a boil. Extract juice. Measure 4 cups into a kettle and add sugar. Follow standard procedure for making jelly with liquid pectin.

STURGEON

One of the things that makes that old classic, *The Swiss Family Robinson,* such a marvelous book is that father Robinson, his wife, and four sons always figure out a way to make use of every plant, animal, fish, etc. that they encounter on their desert island. The description of what they did with the giant sturgeon they caught one day is an example:

> After resting awhile from these exertions, we commenced cleaning and cutting up the fish into pieces to be salted. . . .
>
> "We have such a large supply of fish now," said my wife, "that I think it would be advisable to throw away the roes, the fins, and the tails, as well as the bladders, for they make our dishes and the place smell so unpleasantly."
>
> "Impossible, my dear!" I replied, with a grave face; "from the roes of the sturgeon is prepared that delicate dish named by the Russians 'caviare', and the most excellent glue can be made from the bladder, and those other parts you object to."
>
> My wife shook her head at this information, but as I knew she was right about the unpleasant smell, I at once set to work, that these materials might be got rid of quickly.
>
> The sturgeon's roe, which weighed about thirty pounds, was cleaned and rinsed several times, to remove from it the salt water, and while boiling, the scum of brine was carefully skimmed off, till not a particle remained. The whole mass was

> then placed into a gourd-shell sieve with holes, till the water was pressed entirely from it and then left for twenty-four hours. It was afterward removed from the calabash mould, and mixed with a kind of cheese, made with goat's milk. It only now required to be smoked in the hut, and then removed to our storehouse in the rock, to be preserved as a pleasant and nourishing food for us during the winter.
>
> The fish-bladders next required my attention, as they needed to be separated from the fleshy parts, and also cleaned with fresh water. This done, I cut them into long strips, and after tying a string to each end, stretched them in the sun to dry and become smooth.
>
> The preparation of the fins and tails was really tedious; they had to be skinned, cleaned, and boiled, till they were changed into a perfect jelly, which, after becoming firm, when cool, was thrown into a clean flask, and when thoroughly dried in the sun, and quite hard, was cut into strips and laid by for use. By this process we obtained a supply of really useful glue, which I hoped, when clarified, would not only serve to stick articles together firmly, but also to form a semi-transparent substance to use for window-panes instead of glass.

Sturgeons, giant or otherwise, are no longer plentiful in the United States; and even if they were, we doubt that you would feel compelled to make such extensive use of those you catch. But you should at least cut up the flesh, wrap it in aluminum foil, freeze, and store for 6–9 months.

You should also make caviar. Little special equipment is needed. The most important piece is a rack of 1/4-inch wire mesh beneath which is suspended a piece of insect screening slanted at a 45° angle toward a large metal container.

Open the sturgeon as soon as possible after catching it, remove the roe, and place it (without washing) on the 1/4-inch mesh. Carefully slit the membrane containing the eggs and rub the eggs gently through the mesh onto the screen below. Take care not to break any more eggs than you can help. The eggs roll down the lower screen into the container at the bottom; blood and bits of membrane fall through and are discarded.

When all the eggs are collected, blend salt into them thoroughly with both hands. In cold weather, use 1 lb. dairy salt per 25 lb. roe; in warm weather, use 1 1/2 lb. salt. Mix eggs and salt for 5 to 8 min., until foam develops on top of the mass. Then let the mass stand for 10 min. and mix again until the eggs, now suspended in brine, make a slight but obvious noise as they rub against one another.

Transfer the eggs to a tray with a bottom of 1/32-in. mesh and let them drain for 2 to 4 hr., or until the mass cracks open when you push against the underside of the mesh at one point. Then pour—do not dip—the eggs into a sterilized crock or sterilized Mason jars, place them in a refrigerator, and let them settle for several days. Then add more eggs to fill the containers to capacity; seal tightly and keep in the refrigerator until serving.

SUGAR CANE

Harvest sugar cane as late in the fall as possible, just before a frost. Since the butts are the richest part of the cane, cut the canes close to the ground. Remove leaves and cut off the top ends of the stalks. Crush the stalks to press out the juice. This is not an easy job unless you happen to have a sugar cane mill or a "biscuit break"—in effect, a washing-machine wringer with steel rollers used by southern housewives to make beaten biscuit. But the juice can be extracted by pounding the stalks with the side of an ax.

Strain the juice and let it settle. Then put it in a large kettle and boil it down. Don't stir too much, because this encourages the sirup to turn to sugar, but skim frequently. If you start with 1 gal. juice, it will make about 5 or 6 oz. sirup.

When the sirup reaches the desired consistency (the color will be golden red), remove it from the range and let it cool overnight. The sediment will settle to the bottom. Then in the morning, pour or siphon off the clear sirup into another kettle and reheat to 190°. Pour at once into hot, sterilized jars or bottles and seal. Store in a cool, dark place.

SUGAR MAPLE

Several years ago, shortly after we tapped our sugar maples for the first time, Stan came down with a mysterious discomfort in his chest and was sidelined by the doctor. (The ailment turned out to be minor.) Rather than halt our new sugaring operation, we asked the tenant in our cottage if we could pay him to bring in the sap every day.

His name is Joli. A rather recent emigrant from Italy, he speaks labored English but has become as American as apple pie. He works as a maintenance man in the school system and is delighted to pick up odd jobs that come his way. This time, however, he was clearly uninterested in our proposition though out of kindness he agreed to help.

The reason for his attitude became evident when Stan took him out in the meadow to explain what had to be done. "What you want to do all that for?" Joli finally asked. "You can buy good sirup at the store."

Not everyone, it seems, likes to preserve the fruits of the earth.

Although the sugar maple is the major source of maple sirup and maple sugar, black, silver, and red maples can also be tapped. Tapping is done when the sap begins to run in late winter. In Lyme, we can start in late February; in Vermont, March 1 is a good average starting date. The rate of flow varies from day to day. It is especially plentiful on a warm day following a night of below-freezing temperature. Once a tree starts giving off sap, it will usually continue for several weeks or even a month.

To tap a tree, we drill a 1/2-in. hole slanting slightly upward into the trunk to a depth of about 1 1/2 in. The hole is made on the south side of the trunk at a height of about 3 1/2 ft. Then we drive into the hole a foot-long piece of copper plumbing tube. At the tree end, we cut off the top side of the pipe to catch the sap. At the bottom end, we make a slight notch or dent in the pipe to hold a lightweight 5-qt. plastic paint pail. Though it is not easy to keep it in place on windy days, we try to drape a piece of plastic over the top of the bucket to keep out bugs and dirt.

Collect the sap daily; if it is running well, collect it twice a day. Strain it into a large kettle and boil it down to the desired sirup consistency. This is a slow process because it takes 32 oz. sap to make 1 oz. thin sirup. And since a great deal of moisture is given off, you'll be smart to do the boiling outdoors or in a building where the steam cannot wreck wallpaper and furnishings.

When completed, let the sirup stand overnight or longer to let sediment settle. Then carefully pour or siphon off the clear sirup into sterilized bottles or jars; seal and store in a cool, dark place.

To make maple sugar, simply boil the sap until the temperature reaches 234°. Stir frequently. Strain into hot, sterilized jars or tins and seal.

SUNFISH

There are lots of sunfish and they are known by lots of names, such as sunnies, pumpkinseeds, gillies, crappies, bream, and bluegills. All are small, fresh-water fish, easy to catch, and excellent for pan-frying. Wrap them in aluminum foil and freeze. Store for 6–9 mo.

SUNFLOWERS

Since the birds will do their best to harvest your sunflower seeds before you get to them, it's a good idea to cover the flowers with cheesecloth as the seeds start to mature. Then let the seeds dry thoroughly in the flowers before collecting them in tight jars or plastic bags. Store at room temperature. The seeds will keep longer if you do not roast them until you are ready to use them.

SURINAM CHERRIES

Also called pitanga, the surinam cherry is a tropical evergreen shrub producing small, round, red, or black fruits with prominent ribs and "rabbit ears" at the blossom end. These are juicy and aromatic and range from sweet to slightly acid.

FREEZING

Wash and cut out pits and rabbit ears. Halve fruits and remove seeds. Mix 4 to 5 lb. fruit with 1 lb. sugar or cover the fruit with 40 to 50 percent sugar sirup. Package and freeze. Store for 10–12 mo. Use for pies.

JELLY

Makes 3 8-oz. glasses

5 lb. surinam cherries
4 cups sugar

Wash fruit and remove stem and blossom ends. Put in kettle and mash; then barely cover with water and simmer until soft—about 20 min. Extract juice. Measure 4 cups into a kettle and add sugar. Follow standard procedure for making jelly without added pectin.

SWEET FENNEL

This perennial herb is usually grown as an annual. It has feathery, fernlike leaves with a strong but pleasant licorice flavor. They are used to flavor salads, soups, and fish. Freeze small quantities without blanching.

SWEET MARJORAM

Also known simply as marjoram, this is a small perennial plant usually grown as an annual. The rounded, light green leaves are used in stuffings and sausages, with meats, poultry, and vegetables. If you keep cutting leaves from the top of the plant, it will not flower. Use some of the leaves green. Dry the rest in a warm, airy room and package in bottles or tight polyethylene bags.

SWEET POTATOES

In days gone by, the only sweet potatoes usually to be found in northern grocery stores were mealy, dry-fleshed things. It was a rare occasion when the larger, succulent, orange-fleshed tubers, which southerners call yams, were available.

Happily, this situation has changed considerably. The delicious yam has largely taken over in all stores everywhere.

DRY STORAGE

If you live in a mild or warm climate, you can grow sweet potatoes without too much trouble; but you should cook them rather soon after harvesting, because in order to store the tubers, you must cure them. This is a pretty complicated process requiring very high temperature and humidity.

On the other hand, if you buy sweet potatoes in the store, you can be pretty sure they have already been cured. You should be able to store them for 4–5 mo. if you keep them in a dark place at 55° to 60° and with a high (85 percent) humidity.

FREEZING

Use jumbo potatoes for mashing; large and medium sizes for slicing; small sizes for baking. All must be cured. Bake in a 375° oven until almost tender, or boil in water. Cool completely under a fan. Peel. Leave whole, slice, or mash. To prevent darkening during storage, dip whole sweet potatoes and slices in 1/2 cup lemon juice mixed with 1 qt. water. Mix each quart mashed potatoes with 2 tbs. lemon juice. Pack in rigid containers, seal, and freeze. Store for 10–12 mo. To serve, bake sliced and whole potatoes in a 400° oven for about 30 min. or boil in water till soft. Heat mashed potatoes in the top of a double boiler till soft.

CANNING

Wash sweet potatoes. Boil for about 25 min. until partially soft. Skin and cut in pieces as desired. Pack tightly in jars to 1 in. of top. Press, but not too hard, to fill air spaces. Seal. Process in pressure canner at

10 lb. pressure. Pints for 65 min.; quarts for 95 min.

The alternate method is to wash sweet potatoes and boil them until you can easily remove the skins. Don't overcook. Cut in pieces. Pack in jars to 1 in. of top. Add 1/2 tsp. salt to pints; 1 tsp. to quarts. Cover with boiling water to 1 in. of top. Seal. Process in pressure canner at 10 lb. pressure. Pints for 55 min.; quarts for 90 min.

SWISS CHARD

This is an easily grown relative of the beet. Its big leaves can be boiled or steamed and eaten as a vegetable. To preserve, see *Greens.*

SWORDFISH

Despite all the hullabaloo about swordfish being dangerous to eat because of its supposedly dangerous mercury content, it is still available in many fish stores, and a lot of sensible people are still enjoying it for dinner. If you care to go along with them, have your fish store cut the fish into steaks; wrap them in aluminum foil and put them in the freezer. Store for 6–9 mo.

TAMPALA

This is a leafy vegetable sometimes used as a substitute for spinach. Harvest the entire plants when 6 in. high and freeze or can the leaves like greens (which see).

TANGELOS

Tangelos are crosses between mandarins and grapefruit. They have fruits the size of oranges. For how to store fresh fruit, see initial comment under *Oranges.* You can also freeze tangelos by peeling and sectioning. Pack in rigid containers with 30 percent sugar sirup. Seal and freeze. Store for 10–12 mo.

TANGORS

Tangors are crosses between a tangerine and a sweet orange. They have sweet, juicy, red-orange flesh and peel quite easily. Put them up like oranges.

TARRAGON

There are two tarragons—French and Russian. The latter is easily grown from seed, but the leaves have so little flavor that the plant isn't worth bothering about. The French tarragon is the one to use. Its leaves have an aniselike flavor which does lovely things for fish, poultry, meat, salads, etc. But unfortunately the plant does not set seed, so you must find some one who has plants for sale or to give away.

The best leaves for use in cooking are those put out in the spring and early summer when the plant is about a foot tall. Cut some, not all, of these at that time; dry them in a warm, airy room, and package in tight containers. A second cutting can be made in late summer and preserved in the same way.

To make tarragon vinegar, add 1 cup crushed fresh tarragon leaves to 1 cup wine vinegar; cover; let stand for 2 wk.; then strain into a bottle.

TAUTOG

A black fish found in the North Atlantic, the tautog is very good frozen. Store it for 6–9 mo. It can also be salted.

TEAL

Teal are pint-sized wild ducks and should be handled and frozen in the same way.

TERRAPIN

Terrapins are small water turtles found from the Atlantic Ocean to the Missouri River. They have long been prized for their meat—especially the diamond-back terrapin. Since the creatures usually draw in their heads when you approach, they are easy and safe to catch by hand. But you can also tie a stout fishhook to the end of a short wire attached to a sizable chunk of wood; bait the hook with red meat; and let the lure sail out across the water.

To prepare a terrapin for freezing, scrub it well and drop it, alive, into boiling water. Boil for 10 min.; remove; cool in cold water until you can handle it, and scrub all the skin from the legs, tail, and head with a brush. Then put the terrapin in clean salted water and boil for 1 hr. or a little more. Add some slices of onion, carrot, and celery if you wish. Remove from water and allow to cool. Pull nails from feet. Turn the terrapin on its back and cut away the bottom shell. Remove innards, being especially careful to get out the gall bladder. Save and eat the liver and any eggs you find. Remove the meat and cut it into 1–2-in. pieces. Cut out the bones or not, as you wish.

Cool the meat thoroughly and quickly in the refrigerator. Then package it in rigid plastic containers and freeze. Store for 3–4 mo.

THYME

Of the many varieties of this small perennial plant, the best for seasoning meats, poultry, dressings, soup, stews, cheese, and egg dishes, and salads and vegetables are French thyme and lemon thyme. Cut leafy stems from the plants and dry them completely before attempting to remove the tiny leaves. Store the leaves in tight containers.

TOMATOES

You can put up tomatoes in more different ways than any other vegetable; and it's a good thing you can because when they start flooding in from the garden, they quickly inundate the house. This is especially true if you plant "determinate" varieties. Unlike old-style "indeterminate" tomatoes, these new varieties ripen most of their fruit at the same time. This is fine for farmers who want to get their crops picked and out of the way, but not for the ordinary family.

SKINNING TOMATOES

This is a very simple process but necessary in many preservation methods. Simply wash the tomatoes and then roll around in boiling water for about 30 sec. Dip in cold water. Cut out stem ends and green spots and zip off the skins.

CANNING

Raw pack. Use ripe, sound tomatoes—whole or cut in halves. Pack skinned tomatoes in jars to 1/2 in. of top. Press down to fill air spaces. Add 1/2 tsp. salt to pints; 1 tsp. to quarts. do not add water. Seal. Process in boiling water. Pints for 35 min.; quarts for 45 min.

Hot pack. Cut skinned tomatoes in quarters. Place in kettle and bring to a boil. Stir constantly to prevent sticking. Pack in jars to 1/2 in. of top. Add 1/2 tsp. salt to pints; 1 tsp. to quarts. Seal. Process in boiling water, both pints and quarts, for 10 min.

Tomato juice. Wash tomatoes, remove stem ends and green spots, but don't peel. Cut in pieces. Place in kettle and simmer till soft. Stir frequently. Put through a strainer or food mill. Measure juice and add 1 tsp. salt to each quart. Reheat to boiling and pour into jars to 1/2 in. of top. Seal and process both pints and quarts for 10 min.

FREEZING

Stewed tomatoes. Wash tomatoes, remove stem ends, green spots, and skins. Cut in quarters. Cook covered until tender —about 10 min. Stir enough to prevent sticking but not enough to mash fruit. Set kettle in cold water until tomatoes are completely chilled. Pack in rigid containers, seal, and freeze. Store for 10–12 mo. To serve, heat in double boiler until steaming.

Tomato juice. Wash tomatoes, remove stem ends and green spots, but not skins. Cut in pieces. Simmer in kettle till soft. Stir often. Press through a food mill. Season with 1 tsp. salt per quart. Let cool thoroughly in the refrigerator. Then pour into rigid containers and freeze. To serve, let thaw on counter. Store for 10–12 mo.

RATATOUILLE

An excellent, easy-to-freeze vegetable dish. See recipe under *Eggplant.*

CATSUP

Makes 4 pints

18 lb. tomatoes
1 tbs. peppercorns
1 tbs. mustard seed
1 tsp. thyme
1 tsp. celery seed
1 tsp. oregano
4 bay leaves
1 tbs. dry mustard
3 tbs. salt
2/3 cup sugar
1 tbs. paprika
5/8 tsp. cayenne pepper
2 cups cider vinegar

Wash ripe tomatoes and remove stem ends and green spots, but not skins. Cut in pieces and cook, stirring frequently, till soft. Put through a food mill, or purée in a blender and strain. Tie peppercorns, mustard seed, thyme, celery seed, oregano, and bay leaves in a bag and place in kettle with tomato liquid. Blend the mustard with a little tomato liquid and add to kettle along with salt, sugar, paprika, and cayenne pepper. Cook till thick—about 90 min. Add vinegar during last 10 min of cooking. Remove spice bag. Pour into hot, sterilized jars and seal.

CHILI SAUCE

Makes 5 half pints

8 lb. tomatoes
1 green pepper
4 medium onions
1 qt. cider vinegar
2/3 cup sugar
2 tbs. salt
1/4 tsp. cayenne pepper
2 tsp. ground cloves
2 tsp. ground cinnamon
2 tsp. ground allspice
2 tsp. ground nutmeg

Use ripe tomatoes. Peel, remove stem ends and green spots and cut tomatoes into small pieces to make 2 1/2 qt. Chop pepper and onions fine. Combine all ingredients in a large kettle, bring to a boil, and simmer till thick—about 1 hr. Pour into hot, sterilized jars and seal.

TOMATO PASTE

Makes 9 half pints

25 lb. tomatoes
3 sweet red peppers
2 bay leaves
1 tbs. salt
1 garlic clove

Peel ripe tomatoes, remove stem ends and green spots and cut into small pieces to make 8 qt. Chop peppers to make 1 1/2 cups. Place in large kettle with bay leaves and salt, and simmer for 1 hr. Press through sieve or food mill. Put garlic clove in a bag, add to tomatoes, and continue cooking until thick enough to form a mound on a plate—about 2 1/2 hr. Stir frequently. Lift out garlic. Pour paste into jars to 1/2 in. of top. Seal. Process in boiling water for 45 min.

PICCALILLI

Makes 5 pints

6 lb. green tomatoes
1 large green pepper
1 red chile pepper
1/2 lb. onions
1 cup salt
6 cups cider vinegar
2 cups sugar
1/2 tsp. ground ginger
1/2 tsp. ground cinnamon
1 tbs. mustard seed
1/2 cup horse radish

Wash tomatoes, remove stem ends and chop into coarse pieces. Remove stems and seeds from peppers and chop coarsely.

Slice onions. Combine vegetables, add salt, cover with water, and let stand overnight. Then drain completely. Combine vinegar, sugar, ginger, cinnamon and mustard seed. Add to tomatoes and simmer until tender—about 15 min. Mix in horse radish. Pack at once into hot, sterilized jars and seal.

GREEN TOMATO RELISH

Makes 8 pints

6 lb. green tomatoes
10 sweet red peppers
6 green peppers
3 very large onions
2 tbs. mustard seed
1 tbs. celery seed
2 tbs. salt
1 qt. cider vinegar
4 cups sugar
1 tbs. whole cloves
2 whole allspice
2 sticks cinnamon

Wash and stem tomatoes. Remove stems and seeds from peppers. Peel onions. Then put all the vegetables through a coarse grinder and boil for 15 min. Drain well. Add mustard seed, celery seed, and salt. Meanwhile, in another kettle, boil vinegar, sugar, cloves, allspice, and cinnamon for 20 min. Pour this mixture through a strainer into the vegetable kettle. Bring to a boil. Ladle into hot, sterilized jars and seal.

DILLED GREEN TOMATOES

Makes 9–10 quarts

3/4 cup mixed pickling spices
30 heads green or dry dill weed or 10 tbs. dill seed
20 lb. medium green tomatoes
1 3/4 cups salt
2 1/2 cups cider vinegar
2 1/2 gal. water

Place half the spices and a layer of dill in a 5-gal. crock. Fill to within 3 in. of top with washed, stemmed tomatoes. Cover with half the spice and a layer of dill. Mix salt, vinegar, and water and pour over the tomatoes. Let the tomatoes ferment according to directions in Chapter 13. When the tomatoes turn olive green and are translucent inside, they are ready for packing.

Pack into clean, hot quart Mason jars, but don't wedge too tightly. Add several pieces of dill from the crock. Strain the brine through cheesecloth, heat to boiling, and pour over the tomatoes to 1/2 in. of top. Seal. Process in boiling water for 15 min. Start timing as soon as jars are placed in boiling water.

SPICED TOMATO RELISH

Makes 8 8-oz. glasses

7 lb. tomatoes
1 pt. cider vinegar
6 cups sugar
1 tbs. whole cloves
3 pieces stick cinnamon

Wash and peel ripe tomatoes, remove stem ends and green spots. Cut into small pieces and boil, stirring frequently, for 45 min. Strain off liquid. To pulp and seeds, add vinegar, sugar, and spices tied in a bag. Boil the mixture to thickness of chili sauce

—about 1 hr. Pour into hot, sterilized glasses and seal.

TOMATO CONSERVE

Makes 4 8-oz. glasses

8 tomatoes
3 cups sugar
1/2 cup chopped candied ginger
2 tbs. lemon juice

Use medium-ripe tomatoes. Wash, peel, stem, and chop them. Combine with other ingredients and simmer, stirring often, until mixture is thick enough to form a mound in a spoon. Then ladle into hot, sterilized glasses and seal.

TROUT

One of the sorriest lessons two teenagers ever learned is that if you want to keep fish in condition to eat, you must keep them cool from the time you catch them. It happened in the early thirties when Stan and Jack Pickett hiked into the then remote back country of Yosemite National Park; came upon a pretty little lake, and found it teeming with small golden trout just begging to be caught. The boys did their job; cleaned their catch well; and completely overlooked the fact that on those high, more or less treeless peaks the sun is hot and can build up considerable heat in canvas knapsacks.

When they finally reached their campsite in the valley—all grins with their prowess—they found that the bones were beginning to peel away from the flesh of their beautiful trout. So into the dump went the fish.

FREEZING

The success of commercially frozen trout is ample evidence that freezing is an excellent way to preserve this wonderful fish. Just clean thoroughly; cut off heads, as you wish. Wrap tightly in aluminum foil or package a couple of small trout together in a polyethylene bag. Freeze. Store for 5–7 mo.

CANNING

Clean and cut off fins, tails, and head. Wash thoroughly. Leave trout whole if small; cut larger fish in slices and lengths to fit vertically into pint Mason jars. Leave 1/2 in. head space. Add 1/2 tsp. salt. Seal. Process in pressure canner at 10 lb. pressure for 100 min.

SMOKING

See Chapter 4.

TUNA

FREEZING

Wrap manageable pieces in aluminum foil and freeze. Store for 4–6 mo.

CANNING

Clean and scale fish and remove head and fins. Place belly down on a rack in a large baking pan and bake at 250° until the internal temperature of the fish reaches 170°. It is then cooked through. Let it cool at room temperature and put it in the refrigerator for 12 hr. or longer to firm the meat.

Remove skin, blood vessels at the surface, and discolored flesh. Split fish in half down the backbone and cut the bone out entirely. Remove all other bones including fin bones and all dark brown flesh. Cut white meat remaining crosswise. The width of the pieces should equal the height of the jars you use. Pack solidly into clean jars but don't crush. Leave 3/4 in. head space. To each half pint add 5/8 tsp. salt and 4 tbs. vegetable oil. To each quart, add 1 1/4 tsp. salt and 6 tbs. oil.

Seal jars and process in a pressure canner at 10 lb. pressure. Half pints for 90 min.; pints for 100 min.

TURKEY

Handle domestic and wild turkeys in the same way but allow the latter to age 2 or 3 days in a cold place before processing. Pick birds dry if possible.

FREEZING

Follow standard freezing procedure for poultry. Turkeys may be frozen whole, in halves, quarters, or cut up. Take care to wrap as tightly as possible in order to expel air from the package. Store for 8–10 mo.

CANNING

Turkey to be canned should be boned simply because the bones would take up too much space in the jars. Remove excess fat but not skin. Follow standard canning procedure for meat and poultry. Process in pressure canner at 10 lb. pressure. Pints for 75 min.; quarts for 90 min.

SMOKING

Draw bird completely and wash. Make a brine in the proportions of 8 lb. salt, 3 lb. white sugar, 3 oz. saltpeter, and 6 gal. boiled water. Let this mixture cool completely before using. The saltpeter gives the cured turkey meat a pink color, but it can be omitted.

Place the turkey in a large clean crock and cover completely with the brine. Weight it down if necessary. Put the crock in a refrigerator and let the turkey cure for 1 1/2 days for each pound of dressed weight. If more than one turkey is cured in the crock at the same time, overhaul the pack every week.

When curing is completed, wash the turkey in fresh water, dry thoroughly, and hang it by its wings in the smokehouse. Smoke at 80° to 90° for 24 hr. or at 140° for 5 to 7 hr. until the turkey acquires an amber color. Then cool thoroughly as quickly as possible; wrap in aluminum foil and freeze. Store for 6–8 wk.

TURNIPS

DRY STORAGE

The best variety for storing is Purple-Top White Globe. Cut off tops about 1/4 in. above the bulbs. Keep in damp sand in root cellar. If temperature is close to freezing, the turnips will keep 4–5 mo.

FREEZING

Use small, mild, tender bulbs. Cut off tops. Wash bulbs, and peel and dice in 1/2-in. cubes. Follow standard freezing procedure. Blanch for 90 sec. Store for 10–12 mo. To serve, cook for 12–15 min.

TURNIP GREENS

Process like greens (which see). The best varieties to use for greens are Shogoin, Seven Top, and Just Right.

TURTLE, SNAPPING

Snapping turtles are big, vicious brutes that inhabit fresh-water ponds, rivers, streams, and marshes from the East into the Middle West. The meat is not considered as delectable as that of the terrapin, but is good anyway.

Snapping turtles are usually caught by jug fishing. A large hook baited with red meat is tied to a short length of wire which, in turn, is tied to a sizable chunk of buoyant wood. The wood should be painted red or orange so that you can see it easily in the weeds. Let the lure sail out over the turtles' hunting grounds, and if you're lucky, they will hook themselves.

Since snappers are mean fellows, it's a good idea to shoot or chop their heads off with an ax. Hang them by their tails to drain. Then scrub thoroughly, cut out the bottom shell with a knife, take out the entrails, rinse the body cavity, put in a kettle, cover with water, and boil until the meat is easily cut from the bones. Cool. Remove the upper shell, pull out the claws, and scrub the skin from the feet, head, and tail. Remove the bones or not. Cut the meat into 1–2-in. chunks and put them in a rigid plastic container and freeze. Store for 3–4 mo.

VEAL

Veal can be preserved in the same ways as beef, but because of the delicacy of the meat, freezing is the best way of handling. Store frozen veal for 6–8 mo.

VENISON

Our old golden retriever Juno had an insatiable appetite, and she had the bad habit of crossing Cat Rock Road, where we used to live, and making off with the breakfast dish the Hartleys put out for their dog. But one autumn day when Juno turned up with the hind leg of a deer, we didn't connect it with the Hartleys. And during the following three or four days, when she came back with assorted other pieces of venison, we didn't associate them with the Hartleys either. Then, somehow, the truth came to light.

Having gotten his driver's license, Monty Hartley and a pal went off to the Adirondacks to fulfill a lifetime ambition—to get themselves a couple of deer. They succeeded. And on their return, they hung the animals on a line behind the Hartley house to age. Unfortunately, they failed to consider Juno and the other neighborhood dogs and they hung the deer too low; so when Monty came home from school the next afternoon, his second lifetime ambition—to eat his own kill—had been spirited away into the woods by the wretched dogs.

Since we are not hunters—only food preservers and eaters—we won't presume to tell you how to handle a deer before preserving it. For that information, we turn instead to an excellent leaflet from the Arkansas Agricultural Extension Service. It makes seven quick points:

Bleed your deer—but first be sure he's dead. Shoot again through the jugular vein, halfway between ear and throat.

Stick the deer by standing in back, or close to the body to keep away from the feet. Insert knife 4 or 5 inches at the base of the neck where it joins the chest. Cut sideways to sever veins. Keep open and free from clots. The more blood drained out, the better the meat.

Dress your deer—promptly and carefully. Roll deer over on its back, rump lower than shoulders, and spread hind legs. Tie one hind foot to a tree if you are alone or have difficulty in keeping deer in position.

Make cut along centerline of belly from back to chest cavity to within about 6 in. of tail. Cut through hide first, then through belly muscle. Avoid puncturing paunch or intestines by holding them back with back of hand and guiding knife between first two fingers, cutting edge up.

Cut through diaphragm (thin muscle separating chest from stomach cavity), reach in chest cavity with knife in right hand, and cut windpipe ahead of lungs. A steady pull with left hand will help to roll out the lungs, heart, liver, paunch, and intestines. Take a small sharp blade and cut around anus and draw it back so as to make it come free with intestines. To aid in removal of digestive tract and other organs, you may wish to split the pelvic bone with an ax. This separation of hind quarters will also help in cooling carcass.

Skin your deer. If you cannot hang the deer on a tree, skin it on the ground. If you wish to save the hide, scrape off excess fat and flesh. Fold flesh side inside and freeze hard until you can take to taxidermist, or salt generously with fine meat salt. Fold with flesh side in and roll. Take to taxidermist as soon as possible.

If head and neck mount is desired, stick in brisket area to bleed animal. Skin only to 2 or 3 in. above shoulder and cut head off without skinning neck and head. Fold flesh side in, freeze or salt, and take to taxidermist.

When the deer is skinned, wipe the body cavity with a damp cloth, using as little water as possible. Wet meat spoils more quickly than dry.

Cool your deer—as quickly and thoroughly as possible. Best way is to hang in shade where there is good circulation of air—either end up. Prop flanks open with sticks, clear opening where you stuck the deer.

Save bloodshot parts. Cut away all meat affected by shot. Soak overnight in weak salt water (2 tbs. salt to each quart of water). Drain, dry, and chop or grind for use in deerburgers or stews.

Age your deer. Hang the carcass in a refrigerated or cold area for at least 3 or 4 days. A week or 10 days may be even better. It's best to cover carcass during the aging to keep the meat from drying out. The meat may turn dark during aging, but this doesn't hurt the venison.

Cut up deer properly. Cutting is not a haphazard operation. To do it most easily, hang the carcass by the hocks or hock tendon. Split lengthwise along the backbone from tail to neck. You can use a meat or carpenter's saw, a meat cleaver or hand ax. Cut between the last two ribs and through the backbone to divide halves into quarters. Cut quarters into steaks, roasts, and stew meat in the proportions desired.

And now let's get down to the business of preserving some of the venison for future use.

FREEZING

Trim off as much fat as possible because it tends to change flavor in storage. Wrap each cut separately in aluminum foil and freeze. Store for 10–12 mo. Any meat that is ground should be kept for no more than 4 mo.

SMOKING

Cure and smoke as in Chapter 4. Cure thin cuts for 10 to 14 days; thick cuts for 25 to 30 days. Smoke for about 2 days—or to suit your taste—at 100° to 120°. Venison which has been dry-cured can be wrapped securely and stored in any cool, dark spot for several months.

JERKY

Cut the meat into long strips not over 1/2 in. thick. The strips should run lengthwise of the grain. Salt heavily and apply pepper to keep off insects. Then hang the strips over wires high off the ground in a sunny spot outdoors, and allow them to dry till stiff and leathery.

An easy way to make smoked jerky is to drop the strips of fresh meat for 50 sec. into a boiling brine made with 2 cups salt and 1 gal. water. Remove, drain, and dry thoroughly with paper towels. Then smoke the strips for 3 to 5 hr., depending on how much smoke flavor you like. If the strips are not completely dry at the end of smoking, dry them further in an open oven at low heat.

Whether smoked or unsmoked, jerky should be kept under refrigeration. It will keep for several months.

CANNING

Remove as much fat as possible and cut venison into pieces of convenient, easily packaged size. Put in a shallow pan with very little water to prevent sticking, cover and cook, stirring occasionally, till medium done. Pack loosely in jars to 1 in. of top. Add salt if desired; 1/2 tsp. per pint; 1 tsp. per quart. Cover with boiling meat juice and add boiling water if necessary. Leave 1 in. head space. Seal. Process in a pressure canner at 10 lb. pressure. Pints for 75 min.; quarts for 90 min.

WALNUTS, BLACK

We give you fair warning: If you are going to harvest and store black walnuts, wear the thickest rubber gloves you can find. The brown juice exuded by the husk is a strong dye which defies every solvent and stain remover we have ever tried, and persists for a week or more.

Harvest the nuts as soon as possible after they start showering out of the trees in large numbers. Those that hang on the trees can be knocked down. Then remove the husks by crushing them under you shoe and picking off with gloved hands. Toss the nuts into a tub of water in order to remove some of the brown juice. Of greater importance, it shows which are good nuts and which are bad, because the former sink to the bottom while the latter float.

After the nuts are allowed to dry for several days, store them in sacks in a cool, dark place.

Crack the shells and pick out the kernels whenever you are feeling exceptionally patient and have lots of time to devote to the project. The best way we have found to do the job is to hold one nut at a time on a rock or anvil and rap it in several places with a hammer. If you're lucky, the shell will break up into pieces while the nut meat stays intact. But don't be unhappy if the meat breaks into little pieces too. That's the usual situation.

You can freeze or can the meats in the way described in Chapter 8.

WALNUTS, CARPATHIAN

The Carpathian walnut is a new-to-America variety of Persian walnut and can be grown in cold climates. The nuts are similar to Persian walnuts and are harvested and preserved in the same way.

WALNUTS, PERSIAN

The Persian walnut is more commonly known as the English walnut. It is widely grown only on the West Coast, because it requires a very mild climate. The nuts have much thinner, smoother shells than the black walnut, and the kernels are light brown, mild, and less oily than the black walnut.

When mature, the nuts usually drop out of their hulls onto the ground, but sometimes the hulls adhere and must be removed by hand. In either case, harvest the nuts daily so that they have little chance to deteriorate. Dry them as soon as possible thereafter and let them dry until the kernels snap when broken.

Drying is done by slowly circulating air heated to between 90° and 100° around the nuts. Probably the simplest way to do this is to place the nuts in trays with wire bottoms and direct an electric fan-heater at them. Shake the trays occasionally to turn the nuts so that they are heated more or less evenly on all sides.

In commercial processing the nut shells are bleached after drying. This is not necessary except for the sake of appearance. To bleach nuts at home, soak them for 20 min. in a solution of 1 part household chlorine bleach to 20 parts water. Then rinse in fresh water, and put the nuts in your dryer again for a short time. The nuts are then ready for storing in sacks in a cool, dry place.

Shelled nuts can be frozen or canned.

WALNUTS, SIEBOLD

The Siebold walnut is also known as the Japanese walnut. One of its best varieties is the heartnut. The trees grow mainly in the northeastern quarter of the country, but they are not very popular because the meats are small.

Like all walnuts, the nuts should be harvested as soon as possible after they fall. Remove the husks, wash, and spread the nuts out to dry. Store in sacks or put shelled nuts in the freezer. They can also be canned.

WATERCRESS

Watercress is usually served as a garnish or eaten in salads, but it can also be boiled and eaten like other greens. For how to preserve by freezing or canning, see *Greens.* Use the same blanching and cooking time as for spinach.

One note of caution: If you collect watercress from streams, make sure the water in which it is growing is not polluted.

WATERMELONS

FREEZING

Cut watermelon in half, remove seeds, and scoop out balls; or pare the watermelon and cut the flesh into 3/4-in. cubes. Pack in rigid containers to 1/2 in. of top and cover with 30 percent sugar sirup. Seal and freeze. Store for 10–12 mo.

WATERMELON RIND PICKLES

Makes 4–5 pints

1/2 large watermelon
3/4 cup salt
water
2 trays ice cubes
9 cups sugar
3 cups white vinegar
1 tbs. whole cloves
6 pieces stick cinnamon
1 lemon

Remove green rind and pink meat from watermelon and cut the white rind into 1-in. squares or any shape you like. Make 12 cups. Dissolve salt in 3 qt. water, pour over rind, and add ice cubes. Let stand for 6 hr. Drain and rinse in cold running water. Then put in kettle, cover with cold water, and boil until rind is fork tender, no more —about 10 min. Drain.

Combine sugar, vinegar, and 3 cups water and add spices tied in a bag. Boil for 5 min. Pour over rind. Add spices still in bag. Slice lemon into thin slices remove seeds, and add to sirup. Let stand overnight. In the morning, bring to a boil and simmer until rind is translucent—about 10 min. Then pack rind loosely into jars. Add 1 piece cinnamon from the spice bag to each jar. Fill jars to 1/2 in. of top with boiling sirup. Seal. Process in boiling water for 5 min. Count from time that water in canner returns to a boil.

WEAKFISH

Another large, excellent fish from the Atlantic Ocean, the weakfish is wrapped and frozen in the usual way. Store for 6–9 mo.

WHALE

Chances are you will never catch a whale; and the chances are about equally good that you won't find anyone who has whale meat for sale. But just in case—

Whale meat is surprisingly good. It is red meat, looks like beef steak, and has the same texture. The only way you can tell it's from a sea animal is that it has a fishy taste.

Cut whale meat into steaks or smaller pieces. Wrap in aluminum foil and freeze. Store for 6–9 mo.

WHEAT

Harvest, thresh, and winnow the grain. Then put it through a grist mill to make whole wheat flour. This does not keep as well as white flour, and should ideally be stored in the refrigerator in tight containers.

WHITEFISH

Lake Superior whitefish—there's something to get nostalgic about. It reminds us of crack railroad trains racing through the night, the wheels going clickety-clickety-clickety under the berth, the whistle wailing at the crossroads, the bell clanging as we glided through the yards to stations bright with light. Most of all, whitefish reminds us of dining cars with white-linen-covered tables, polished silverware, sparkling glassware, smiling, always friendly black waiters.

Lake Superior whitefish was often on the menus of the old diners, and it was something that travelers who loved food always ordered. It was that good.

Unhappily, the whitefish population has declined with the railroads. But if you are fortunate enough to catch or find any in the fish markets, by all means grab it. Not just enough for dinner tonight but for future use. It keeps for 6–9 mo. in the freezer if you wrap it tightly in aluminum foil.

Whitefish also makes an excellent smoked product when handled according to directions in Chapter 4. It can also be salted.

Two other fishes closely related to whitefish and put up in the same ways are grayling and the lake herring, also called cisco.

WHITING

Whiting, also called silver hake, is one of several hakes that are caught for food, but it is the most widely available. A salt-water fish, it averages a couple of pounds. Wrap and freeze in the usual way. Store for 7–10 mo.

To smoke whiting, follow directions in Chapter 4 for cold smoking.

WILD RICE

Wild rice is a tall grass growing in shallow fresh waters in many parts of the country, but mainly in the Great Lakes region. It must be harvested just as it matures, because the spikes shatter soon after that. The easiest way to gather the grain is from a boat. Simply pull the spikes down and strike them with a stick till the grains fall into the boat.

After hulling, store the rice in tight containers. It is normally boiled and eaten whole, especially with duck. It may also be ground into flour in a grist mill.

WINEBERRIES

Wineberries are a close relative of the raspberry and have small red, raspberrylike fruits. The plants grow in the East. You can freeze the berries and make them into jam and jelly just as you handle raspberries.

WINTER SAVORY

This is a perennial very similar to summer savory. The leaves can be dried, stored, and used in the same way.

WOLFFISH

The wolffish is a mean-looking salt-water fish with sizable tusks that can make a nasty hole in your hand if you don't watch out. So take care when pulling the fish in. Wrap and freeze them in aluminum foil. Store for 6–9 mo.

WOODCOCK

Draw woodcock as soon as possible after shooting. Let cool rapidly. Remove feathers by dry-plucking. Cut off neck, also legs and wings at the second joint. Age in the refrigerator for at least 24 hr. Wrap in aluminum foil or put in polyethylene bags and freeze. Store for 9 mo.

YOUNGBERRIES

A variety of blackberry, the youngberry produces large berries with only a few seeds. They can be put up in the same ways as blackberries.